Math Contests
Grades 7 and 8
(and Algebra Course 1)
Volume 4

**School Years
1996-1997 through 2000-2001**

Written by

Steven R. Conrad • Daniel Flegler

Published by MATH LEAGUE PRESS
Printed in the United States of America

Cover art by Bob DeRosa

Phil Frank Cartoons Copyright © 1993 by CMS

First Printing, 2001

Copyright © 2001
by Mathematics Leagues Inc.
All Rights Reserved

Math League Press
P.O. Box 17
Tenafly, NJ 07670-0017

ISBN 0-940805-13-8

Preface

Math Contests—Grades 7 and 8, Volume 4 is the fourth volume in our series of problem books for grades 7 and 8. The first three volumes contain the contests given in the school years 1979-1980 through 1995-1996. This volume contains the contests given from 1996-1997 through 2000-2001. (You can use the order form on page 154 to order any of our 12 books.)

This book is divided into three sections for ease of use by students and teachers. You'll find the contests in the first section. Each contest consists of 30 or 40 multiple-choice questions that you can do in 30 minutes. On each 3-page contest, the questions on the 1st page are generally straightforward, those on the 2nd page are moderate in difficulty, and those on the 3rd page are more difficult. In the second section of the book, you'll find detailed solutions to all the contest questions. In the third and final section of the book are the letter answers to each contest. In this section, you'll also find rating scales you can use to rate your performance.

Many people prefer to consult the answer section rather than the solution section when first reviewing a contest. We believe that reworking a problem when you know the answer (but *not* the solution) often leads to increased understanding of problem-solving techniques.

Each school year, we sponsor an Annual 7th Grade Mathematics Contest, an Annual 8th Grade Mathematics Contest, and an Annual Algebra Course 1 Mathematics Contest. A student may participate in the contest on her or his grade level or for any higher grade level. For example, students in grade 7 (or below) may participate in the 8th Grade Contest. *Any* student may participate in the Algebra Course 1 Contest. Starting with the 1991-92 school year, students have been permitted to use calculators on any of our contests.

Steven R. Conrad & Daniel Flegler, contest authors

Acknowledgments

For demonstrating the meaning of selflessness on a daily basis, special thanks to Grace Garcia.

To Mark Motyka, we offer our gratitude for his assistance over the years.

To Jeannine Kolbush, who did an awesome proofreading job, thanks!

Table Of Contents

The Contests

· ·

1996-1997 through 2000-2001

7th Grade Contests

1996-1997 through 2000-2001

1996-97 Annual 7th Grade Contest

Tuesday, February 4, 1997

7

Instructions

- **Time** You will have only *30 minutes* working time for this contest. You might be *unable* to finish all 40 questions in the time allowed.

- **Scores** Please remember that *this is a contest, not a test*—and there is no "passing" or "failing" score. Few students score as high as 30 points (75% correct). Students with half that, 15 points, *should be commended!*

- **Format and Point Value** This is a multiple-choice contest. Each answer is an A, B, C, or D. Write each answer in the *Answers* column to the right of each question. A correct answer is worth 1 point. Unanswered questions get no credit. You **may** use a calculator.

Copyright © 1997 by Mathematics Leagues Inc.

	Answers
1. $10 \times (73 + 37) = 10 \times (91 + \underline{\ ?\ })$ A) 9 B) 19 C) 37 D) 110	1.
2. Of the following fractions, which is the largest? A) $\frac{3}{4}$ B) $\frac{5}{8}$ C) $\frac{11}{16}$ D) $\frac{37}{64}$	2.
3. Each of the following has 9 as a factor *except* A) 18 018 B) 18 081 C) 81 180 D) 81 181	3.
4. To the nearest 1%, what percent of the 12 months have 31 days? A) 75% B) 70% C) 58% D) 50%	4.
5. In a *Sing Thing*, the first 4 singers sang Song 1, the next 4 sang Song 2, and so forth. Singer number 73 sang Song $\underline{\ ?\ }$. A) 18 B) 19 C) 20 D) 73	5.
6. 25% of 8 is equal to 50% of A) 64 B) 16 C) 4 D) 2	6.
7. A sheet of paper forms a right triangle when I fold it in half diagonally. The unfolded paper could be a A) square B) triangle C) trapezoid D) pentagon	7.
8. If the value of 10 coins is \$2.20, then 8 of these coins could be A) pennies B) nickels C) dimes D) quarters	8.
9. There are $\underline{\ ?\ }$ times as many seconds in an hour as in a minute. A) 24 B) 60 C) 360 D) 3600	9.
10. Of the following, which is the smallest? A) 1.001×10^4 B) 1.01×10^3 C) 1.01×10^2 D) 1.1×10^2	10.
11. What number is *not* the reciprocal of any number? A) 0 B) 0.5 C) 1 D) 2	11.
12. What is 404.404, rounded to the nearest tenth? A) 400 B) 404.5 C) 404.40 D) 404.4	12.
13. What is 404.404, rounded to the nearest hundredth? A) 400 B) 404.5 C) 404.40 D) 404.4	13.
14. Of the following, which is closest to 0.62? A) $\frac{6}{10}$ B) $\frac{7}{11}$ C) $\frac{10}{17}$ D) $\frac{11}{18}$	14.
15. *Rent-A-Kid* got Pat a babysitting job. Pat's age today, plus Pat's age 5 years ago, equals 21. Pat's age today is A) 12 B) 13 C) 14 D) 15	15.

Go on to the next page ⮕ **7**

16. When the sum of two primes is prime, one of the primes *must* be
    A) 5   B) 3   C) 2   D) 1

16.

17. I'm thinking of three numbers. The product of two of them is 24, and the sum of all three is also 24. The three numbers could be
    A) 3, 8, 13   B) 4, 6, 16   C) 2, 12, 12   D) 1, 24, 1

17.

18. The average degree-measure of an angle of a triangle is
    A) 45°   B) 60°   C) 90°   D) 180°

18.

19. What tax rate will result in a 30¢ tax on a $5 purchase?
    A) 5%   B) 6%   C) 8%   D) 30%

19.

20. In Volume Two of a series, the first page is numbered 282 and the last page is numbered 562. How many pages are in Volume Two?
    A) 279   B) 280   C) 281   D) 282

20.

21. The dimensions of four rectangles are given below. Which has an area different from the area of each of the other three?
    A) 13×25 B) 7×45 C) 15×21 D) 9×35

21.

22. To celebrate the first day of a leap year, I taught my dog to jump through a hoop. It was a Sunday. When he taught me the same trick the first day of the next year, it was a ? .
    A) Sunday   B) Monday
    C) Tuesday   D) Wednesday

22.

23. A decagon has ? sides.
    A) 5   B) 6   C) 8   D) 10

23.

24. The largest circular shape my piece of string can form has a diameter 8 cm long. How long is the diameter of the largest circular shape that half my piece of string can form?
    A) $4/\pi$ cm   B) $8/\pi$ cm   C) 2 cm   D) 4 cm

24.

25. What is the value of $\frac{3}{4} \times \frac{7}{8} \times \frac{15}{16}$, rounded to the nearest 0.25?
    A) 0.25   B) 0.50   C) 0.75   D) 1.00

25.

26. On Tues., Dec. 3, I began drinking a glass of cola every day except Sat. and Sun. I drank my 22nd glass of cola on
    A) Dec. 24   B) Dec. 25   C) Dec. 31   D) Jan. 1

26.

27. Which is 25% of 100×100?
    A) 25×25 B) 50×50 C) 200×200 D) 400×400

27.

28. The product of ? and its square is 125.
    A) 5   B) 10   C) 25   D) $\sqrt{125}$

28.

29. The largest odd factor of $6^8 \times 10^6$ is
    A) $(6^8 \times 10^6)/2$ B) $3^4 \times 5^3$ C) 15^6 D) $3^8 \times 5^6$

29.

Go on to the next page ▐▐▐➡ **7**

7

30. The hypotenuses of four congruent isosceles right triangles serve as the sides of a square, as shown. If the area of that (unshaded) square is 18, what is the area of the shaded region?

A) 36 B) 27 C) 18 D) 9

30.

31. If the days of the year were numbered backward using consecutive numbers (so that December 31 is day number 1), in what month would day number 123 occur?

A) August B) July C) March D) September

31.

32. Each wheel on Harley's bike has a radius of 60 cm. How many revolutions does each wheel make when Harley bikes 500 m?

A) $\frac{500}{120\pi}$ B) $\frac{1000}{120\pi}$ C) $\frac{50000}{120\pi}$ D) $\frac{100000}{120\pi}$

32.

33. 5 consecutive whole numbers *cannot* add up to

A) 225 B) 222 C) 220 D) 200

33.

34. Bo, Jo, Mo, and Di went to see a hit movie. In how many ways can they line up, one behind another, waiting to see the movie?

A) 4 B) 6 C) 16 D) 24

34.

35. If 10^{18} is 1 quintillion, which is half of 1 quintillion?

A) 5^9 B) 5^{18} C) 10^9 D) 5×10^{17}

35.

36. Ann says "I never lie." Bob says "Ann is not lying." Carol says "Bob is lying." Dan says "Carol is not lying." If Bob is lying, how many of the other three must be lying?

A) 0 B) 1 C) 2 D) 3

36.

37. If the hands on a round clock moved *counter-clockwise*, to what number would the minute hand point 20 minutes before the hour hand pointed to the 3?

A) 4 B) 7 C) 8 D) 11

37.

38. Of the following numbers, which has the smallest ones' digit?

A) 2^{53} B) 2^{52} C) 2^{51} D) 2^{50}

38.

39. A cloth tape measure, when folded in half 5 times lengthwise, is 5 cm long. How long was the tape before it was first folded?

A) 25 cm B) 80 cm C) 100 cm D) 160 cm

39.

40. Pat paid 1/8 of the restaurant bill, Jan paid 1/4 of it, and I paid the rest, $1.20. How much was the restaurant bill?

A) $1.80 B) $1.92 C) $2.00 D) $3.20

40.

The end of the contest 🖎 **7**

Solutions on Page 73 • Answers on Page 138

1997-98 Annual 7th Grade Contest

Tuesday, February 3, 1998

Instructions

7

- **Time** You will have only *30 minutes* working time for this contest. You might be *unable* to finish all 40 questions in the time allowed.

- **Scores** Please remember that *this is a contest, not a test*—and there is no "passing" or "failing" score. Few students score as high as 30 points (75% correct). Students with half that, 15 points, *should be commended!*

- **Format and Point Value** This is a multiple-choice contest. Each answer is an A, B, C, or D. Write each answer in the *Answers* column to the right of each question. A correct answer is worth 1 point. Unanswered questions get no credit. You **may** use a calculator.

1. $0.77 \div 7 =$
    A) $1 \div 11$ B) $0.66 \div 6$ C) $7 \div 0.77$ D) $77 \div 7$

    1.

2. Every student here has 3 pencils and 2 pens. If they have 1998 more pencils than pens altogether, then there are ? students here.
    A) 1998   B) 3996   C) 5994   D) 9990

    2.

3. Of the following, which equals 13.875?
    A) $\frac{112}{8}$ B) $12\frac{7}{8}$ C) $13\frac{1}{8}$ D) $13\frac{7}{8}$

    3.

4. $\frac{1}{5} + \frac{2}{10} + \frac{3}{15} + \frac{4}{20} + \frac{5}{25} =$
    A) $\frac{1}{5}$ B) $\frac{15}{25}$ C) $\frac{35}{50}$ D) 1

    4.

5. 1998 cm = ? m
    A) 0.1998   B) 1.998   C) 19.98   D) 199.8

    5.

6. If $5 \times \blacklozenge = 5 + 4 + 3 + 2 + 1$, then $\blacklozenge =$
    A) 3   B) 5   C) 10   D) 15

    6.

7. $15^4 =$
    A) $3^2 \times 5^2$ B) 3×5^4 C) $3^3 \times 5^4$ D) $3^4 \times 5^4$

    7.

8. I took my medicine at 8:45 A.M and again, 3.5 hours later, at
    A) 11:15 A.M.   B) 11:15 P.M.
    C) 12:15 A.M.   D) 12:15 P.M.

    8.

9. $1 \times 1 + 2 \times 20 + 3 \times 300 + 4 \times 4000 =$
    A) 2541   B) 4321   C) 16 940   D) 16 941

    9.

10. How many different prime numbers are factors of 1 million?
    A) one   B) two   C) three   D) ten

    10.

11. Of the following, which is *not* divisible by 11?
    A) 111   B) 1111   C) 111 111   D) 11 111 111

    11.

12. Round 19.975 to the nearest hundredth.
    A) 19.97   B) 19.976   C) 19.98   D) 20

    12.

13. In 1997, my town had 180 sunny days, 140 cloudy days, and 45 rainy days. What was the ratio of sunny to non-sunny days?
    A) 140:180   B) 180:140   C) 180:185   D) 180:365

    13.

14. Which of the following is *not* a factor of 1998?
    A) 37   B) 36   C) 27   D) 18

    14.

15. What is the reciprocal of 0.0001?
    A) 0.9999   B) 1.0001   C) 10 000   D) 100 000

    15.

16. If the product of some primes is 15 015, then ? is *not* a factor.
    A) 2   B) 3   C) 5   D) 7

    16.

Go on to the next page ⅠⅠⅠ➡ **7**

10

17. The reciprocal of a positive whole number is always
    A) less than or equal to 1 B) less than 1
    C) greater than 1   D) greater than or equal to 1

18. Ted is 5 years older than Bob, and Ann is 3 years younger than Bob. If the sum of Ted's and Ann's ages is 30, how old is Bob?
    A) 25   B) 20   C) 16   D) 14

19. $\sqrt{\frac{1}{16} + \frac{1}{16} + \frac{1}{16} + \frac{1}{16}}$ =
    A) 1   B) 0.5   C) 0.25   D) 0.125

20. A plain beanie costs 29¢. One with a propeller costs 31¢. Altogether, I spent $3.02 on beanies. I bought _?_ with propellers.
    A) 3   B) 4   C) 5   D) 6

21. Which equals $\frac{2}{3}$ % of 600?
    A) 4   B) 6   C) 40   D) 400

22. Of the following, which has a value different from the others?
    A) $\frac{4^3}{2^9}$ B) $\frac{2^6}{8^3}$ C) $\frac{1}{8}$ D) $\frac{1}{4}$

23. A race is 22.5 km long. If I want to finish the race in 7.5 hours or less, then my average running speed *must* be _?_ or faster.
    A) 2.5 km/hr   B) 3 km/hr   C) 3.5 km/hr   D) 4 km/hr

24. Six students line up with Ann first, Bob next, and Ray as the 3rd student behind Ann. Jan is next to Pat. Carol must be between
    A) Bob & Ray B) Ann & Bob C) Ray & Pat D) Ray & Jan

25. My street's mailman also delivers mail on exactly 6 other streets. If he delivers mail to 210 houses every day, what is the average number of houses on each street of his route?
    A) 30   B) 32   C) 34   D) 35

26. $1997^{1998} \div 1997^{1997}$ =
    A) 1   B) 1997   C) 1998   D) 1997^2

27. The average of 3 consecutive primes, each less than 50, could be
    A) 3   B) 5   C) 7   D) 11

28. I have 5 coins of equal value in my left pocket, and 2 coins of equal value in my right pocket. If the total value of the coins in each pocket is the same, my left pocket contains
    A) pennies B) nickels C) dimes D) quarters

29. Patty placed four 3×5 photos on a page in her photo album, without overlapping. If a page measures 8×12, what is the area of that part of the page not covered by photos?
    A) 27   B) 35   C) 36   D) 81

30. When multiplied out, $999\,999\,999\,999^2$ has _?_ digits. A) 22 B) 23 C) 24 D) 144	30.
31. A ball rolled around the circular rim of a net twice before falling through the net. If the diameter of the rim is 50 cm long, how far did the ball travel around the rim? A) 50π cm B) 100π cm C) 150π cm D) 200π cm	31.
32. George, Paul, and Ringo each thought of a number. The sums of each possible pairing of numbers are 1997, 1998, and 1999. What is the sum of their three numbers? A) 1998 B) 2991 C) 2997 D) 5994	32.
33. All sides of a right triangle are integers. The hypotenuse *could* be A) 10 B) 12 C) 16 D) 18	33.
34. Jane's hair grows at the rate of 0.5 cm every 4 weeks. She cuts off 0.25 cm every 6 weeks, and she just cut it today. Right now, Jane's hair is 50 cm long. Eleven weeks from today, Jane's hair will be _?_ cm long. A) 51 B) 51.125 C) 51.375 D) 51.5	34.
35. If the sum of the squares of two sides of a triangle equals the square of the third side, then the largest angle of this triangle is A) 45° B) 60° C) 90° D) 180°	35.
36. $1.998 \times 10^{20} - 1.997 \times 10^{20} =$ A) 10^{16} B) 10^{17} C) 10^{23} D) 10^{24}	36.
37. If 1 tic = 3 tacs, and 4 tacs = 5 toes, how many tics equal 1 toe? A) $\frac{1}{15}$ B) $\frac{1}{5}$ C) $\frac{4}{15}$ D) $\frac{5}{12}$	37.
38. A ferris wheel ride costs 3 tickets. A carousel ride costs 2 tickets. If I actually spend 30 tickets and go on the two rides at least once each, at *most* how many times can I ride the carousel? A) 15 B) 14 C) 13 D) 12	38.
39. How many different integers have reciprocals whose values are greater than $\frac{19}{1998}$ and less than $\frac{98}{1998}$? A) 78 B) 79 C) 84 D) 85	39.
40. If I add together _?_ different positive odd numbers, the value of the sum can be 12321, but it can never be less than 12321. A) 111 B) 121 C) 131 D) 221	40.

The end of the contest ✍ **7**

Solutions on Page 77 • Answers on Page 139

1998-99 Annual 7th Grade Contest

Tuesday, February 2, 1999

Instructions

7

- **Time** You will have only *30 minutes* working time for this contest. You might be *unable* to finish all 40 questions in the time allowed.

- **Scores** Please remember that *this is a contest, not a test*—and there is no "passing" or "failing" score. Few students score as high as 30 points (75% correct). Students with half that, 15 points, *should be commended!*

- **Format and Point Value** This is a multiple-choice contest. Each answer is an A, B, C, or D. Write each answer in the *Answers* column to the right of each question. A correct answer is worth 1 point. Unanswered questions get no credit. You **may** use a calculator.

1. Of the following numbers, which is closest to 19.99? | 1.

   A) 19   B) 19.9   C) 20   D) 20.01

2. $(2001 - 1000) + (1998 - 1000) = 1999 - \underline{?}$ | 2.

   A) 0   B) 1   C) 1000   D) 2000

3. At the cookout, I ate as many hot dogs as the millionths' digit of π. Since $\pi =$ 3.14159265359 . . . , I ate ? hot dogs. | 3.

   A) 2   B) 5   C) 6   D) 9

4. $\sqrt{1 \times 9 \times 9 \times 9}$ = | 4.

   A) 9   B) 19   C) 27   D) 99

5. $1024 \div 64 \div 8 \div 2$ = | 5.

   A) 1   B) 2   C) 4   D) 16

6. $(2 \text{ m}) - (1000 \text{ mm})$ = | 6.

   A) 10 cm   B) 100 cm   C) 20 cm   D) 200 cm

7. Of the following, which is *not* equal to 1? | 7.

   A) 1% of 100   B) $100 \times 1\%$ C) $1 \times 100\%$ D) 1% of 1

8. Twelve hours and one minute before midnight is | 8.

   A) 12:01 P.M.   B) 12:01 A.M.   C) 11:59 P.M.   D) 11:59 A.M.

9. $6 \times (3 + 15) = (2 + 10) \times \underline{?}$ | 9.

   A) 5   B) 9   C) 11   D) 12

10. Round the quotient $10\,499 \div 1000$ to the nearest 1. | 10.

    A) 1   B) 1.1   C) 10   D) 11

11. 100% of 50 = 200% of ? | 11.

    A) 25   B) 75   C) 100   D) 150

12. $\sqrt{9} + \sqrt{4} + \sqrt{9} + \sqrt{4}$ = | 12.

    A) $\sqrt{10}$ B) $\sqrt{26}$ C) $\sqrt{36}$ D) $\sqrt{100}$

13. At the Founder's Day Gift Wrap Sale, how many gifts can I get wrapped for $2.00 if the charge for wrapping is only 5¢ per gift? | 13.

    A) 10   B) 20   C) 40   D) 100

14. $1\frac{1}{2} \times 1\frac{1}{3} \times 1\frac{1}{4} \times 1\frac{1}{5}$ = | 14.

    A) 2   B) 3   C) 6   D) 12

15. If the diameter of a circle is an integer, its radius *could* be | 15.

    A) 7.25   B) 7.5   C) 7.75   D) π

Go on to the next page ⏩ **7**

16. The largest power of 4 which is a factor of $2^2 \times 3^3 \times 4^4$ is A) 4 B) 4^3 C) 4^4 D) 4^5	16.
17. The announcer just called my number. It's the largest prime factor of 26 000 000 000. What's my number? A) 2 B) 5 C) 13 D) 26	17.
18. 200% of _?_ is a prime. A) 1 B) 2 C) 3 D) 5	18.
19. 5% is what percent of 0.1? A) 10% B) 50% C) 200% D) 500%	19.
20. How many square tiles measuring 2 cm × 2 cm are needed to cover a rectangular area measuring 40 cm × 50 cm? A) 500 B) 1000 C) 2000 D) 8000	20.
21. If the total value of an equal number of pennies, nickels, and dimes is $2.40, what is the value of the nickels alone? A) 45 cents B) 50 cents C) 60 cents D) 75 cents	21.
22. If my hair, now 3 cm long, grows 0.5 cm every day, my hair will be 8 cm long in _?_ days. A) 25 B) 16 C) 10 D) 6	22.
23. My favorite trophy holds 1ℓ of water. After I drink 2/3 of it, then replace 3/4 of what I drank, how many ℓ of water will remain in my trophy? A) $\frac{1}{2}\ell$ B) $\frac{2}{3}\ell$ C) $\frac{3}{4}\ell$ D) $\frac{5}{6}\ell$	23.
24. $7^2+7^2+7^2+7^2+7^2+7^2+7^2 = 7 \times$ _?_ A) 7^2 B) 7^3 C) 7^7 D) 7^8	24.
25. $1-\frac{1}{3}-\frac{1}{6}-\frac{1}{9}-\frac{1}{18} =$ A) $\frac{1}{6}$ B) $\frac{2}{9}$ C) $\frac{1}{3}$ D) $\frac{1}{2}$	25.
26. If 2 quadrilaterals share a side, the resulting figure could *not* be A) a hexagon B) a pentagon C) a square D) a triangle	26.
27. If the horizontal lines are parallel, then $x°+y° =$ 135° A) 80° B) 90° C) 100° D) 110° $x°$ $y°$	27.
28. Of the following fractions, which is smallest? A) $\frac{1}{9}$ B) $\frac{1}{19}$ C) $\frac{19}{199}$ D) $\frac{199}{1999}$	28.

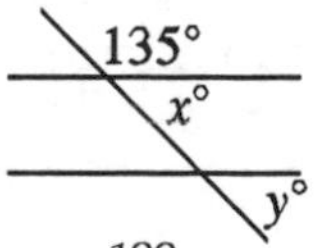

Go on to the next page �{IIII➡} **7**

29. A square of side 3 is cut from a square of side 9 as shown. What is the area of the shaded region?

    A) 6   B) 24   C) 64   D) 72

    29.

30. The least common multiple of (2×3), (3×5), and (2×5) is

    A) 1   B) 20   C) 30   D) 900

    30.

31. Three congruent circles are tangent as shown. If the trian- gle has a perimeter of 30, what is the area of each circle?

    A) 25π B) 36π C) 75π D) 100π

    31.

32. From noon one day till noon 2 days later, how many times will the hour hand of a clock pass the 1 o'clock mark?

    A) 2   B) 3   C) 4   D) 48

    32.

33. $0.4444 + 0.5555 + \underline{\ ?\ } = 1$.

    A) 0.00001   B) 0.0001   C) 0.0111   D) 0.1111

    33.

34. When Weird Uncle Barney came to visit, we told him that the number of days he could stay was the greatest common divisor of 1999 and 2001. How many days is that?

    A) 1   B) 2   C) 3   D) 7

    34.

35. What is the ones' digit of 2^{1999}?

    A) 2   B) 4   C) 6   D) 8

    35.

36. If the angles of a triangle have whole-number degree measures, and one angle is 80°, the largest possible remaining angle is

    A) 79°   B) 80°   C) 99°   D) 100°

    36.

37. Call a point at which two or more lines intersect a *crossing*. Drawing ten different lines *cannot* result in $\underline{\ ?\ }$ different crossings.

    A) 0   B) 1   C) 2   D) 9

    37.

38. An equilateral triangle shares a side with an isos- celes right triangle as shown. What is $m\angle ABC$?

    A) 105°   B) 120°   C) 135°   D) 150°

    38.

39. Choice $\underline{\ ?\ }$ has more different prime factors than the other choices.

    A) 1997   B) 1998   C) 1999   D) 2000

    39.

40. **1,1,1,2,1,3,1,4,1,** What is the 200th number in the sequence 1, 1, 1, 2, 1, 3, 1, 4, 1, 5, . . . ?

    A) 1   B) 100   C) 101   D) 200

    40.

The end of the contest **7**

Solutions on Page 81 • Answers on Page 140

16

1999-2000 Annual 7th Grade Contest

February, 2000

Instructions

7

- **Time** You will have only *30 minutes* working time for this contest. You might be *unable* to finish all 40 questions in the time allowed.

- **Scores** Please remember that *this is a contest, not a test*—and there is no "passing" or "failing" score. Few students score as high as 30 points (75% correct). Students with half that, 15 points, *should be commended!*

- **Format and Point Value** This is a multiple-choice contest. Each answer is an A, B, C, or D. Write each answer in the *Answers* column to the right of each question. A correct answer is worth 1 point. Unanswered questions get no credit. You **may** use a calculator.

Copyright © 2000 by Mathematics Leagues Inc.

1. $111\,111 + 222\,222 + 333\,333 + 444\,444 = 222\,222 \times \underline{\ ?\ }$

   A) 1   B) 4   C) 5   D) 10   1.

2. In which of the following divisions is the remainder equal to 2?

   A) $257 \div 5$ B) $228 \div 6$ C) $195 \div 3$ D) $176 \div 4$ 2.

3. 6 twos + 8 threes = 2 sixes + $\underline{\ ?\ }$ eights

   A) 3   B) 6   C) 8   D) 12   3.

4. $2 + (10 \times 2) + (100 \times 2) + (1000 \times 2) =$

   A) 224   B) 2000   C) 2220   D) 2222   4.

5. Of the following fractions, which represents a whole number?

   A) $\dfrac{182}{7}$ B) $\dfrac{172}{12}$ C) $\dfrac{189}{17}$ D) $\dfrac{178}{21}$ 5.

6. $2^2 + 2^2 + 2^2 + 2^2 = 2^2 \times \underline{\ ?\ }$

   A) 2^1 B) 2^2 C) 2^3 D) 2^4 6.

7. Of the amounts listed below, the one with the largest tens' digit is exactly what we earned on our *Walkathon*. That amount is

   A) \$1231.21   B) \$1123.03   C) \$3010.30   D) \$2302.12   7.

8. It's possible for a February to have $\underline{\ ?\ }$ Tuesdays, but not more.

   A) 3   B) 4   C) 5   D) 6   8.

9. Find the missing number: $\dfrac{1+2}{3} + \dfrac{4+5}{6} = \dfrac{7+8}{9-?}$

   A) 0   B) 3   C) 6   D) 12   9.

10. A rectangle whose width is 3 has the same area as a square whose side is 9. What is the perimeter of this rectangle?

    A) 27   B) 36   C) 60   D) 81   10.

11. The product of 2 different numbers, both greater than 0, must be

    A) greater than 0   B) greater than 1
    C) greater than 2   D) at least 2   11.

12. Ozzie hid his head in the sand 98 hours after 11 P.M. Sunday. Ozzie hid his head on a

    A) Tues.   B) Wed.   C) Thurs.   D) Fri.   12.

13. What is the largest prime factor of $2^2+3^2+5^2$?

    A) 5   B) 13   C) 19   D) 37   13.

14. The reciprocal of the smallest prime is

    A) 0   B) $\dfrac{1}{2}$ C) 1   D) 2   14.

15. $(20 \times 100) - (20 \times 10) - (20 \times 1) =$

    A) 20×111 B) 20×109 C) 20×91 D) 20×89 15.

Go on to the next page ⫸ **7**

16. 4.5 hours is equivalent to each of the following *except* A) $\frac{3}{16}$ day B) 270 min C) 16 200 sec D) $\frac{3}{100}$ week	16.
17. What fraction of 1 m is 15 cm? A) $\frac{1}{10}$ B) $\frac{3}{20}$ C) $\frac{1}{15}$ D) $\frac{10}{15}$	17.
18. How many thousandths, when added together, equal one tenth? A) 100 B) 1000 C) 10 000 D) $\frac{1}{100}$	18.
19. When 250 adults were asked if they could use a hammer, 40% said "Yes!," 38% said "No!," and the rest said "Ouch." How many said "Ouch"? A) 22 B) 25 C) 55 D) 195	19.
20. $\sqrt{25 - 16} =$ A) 9 B) 3 C) 1 D) –11	20.
21. Jack is as old now as Jill was 3 years ago. If the sum of their ages is 43, how old will Jill be in 2 years? A) 20 B) 22 C) 23 D) 25	21.
22. On 5 math tests, I averaged 95. The sum of my 5 test scores was A) 95 B) 100 C) 475 D) 495	22.
23. What is the reciprocal of $\left(1 + \frac{7}{8}\right)$? A) $\frac{8}{15}$ B) $1 + \frac{8}{7}$ C) $\frac{15}{8}$ D) $\frac{7}{15}$	23.
24. Whenever the value of my dimes is one-fifth the value of my quarters, I will have _?_ as many dimes as quarters. A) one-half B) one-fifth C) two-thirds D) twice	24.
25. What is the average number of days per month for the year 2000? A) 29 B) 30 C) 30.5 D) 31	25.
26. How many of the first 10 whole numbers are factors of 162? A) 6 B) 5 C) 4 D) 3	26.
27. Grandpa's age in years equals his dog's age in months. Grandpa is 55 years older than his dog. Grandpa's dog is _?_ months old. A) 48 B) 55 C) 60 D) 66	27.
28. 30 is 0.1% of A) 3 B) 300 C) 3000 D) 30 000	28.
29. What number is midway between 1 234 567 and 7 654 321? A) 3 765 432 B) 4 321 321 C) 4 444 444 D) 3 456 789	29.

Go on to the next page ⮞ **7**

30. What is the area of a square whose perimeter is 3 cm?

    A) $\frac{9}{16}$ cm^2 B) $\frac{3}{2}$ cm^2 C) 3 cm^2 D) 9 cm^2

    30.

31. $34\,592\,867\,544^2 - 34\,592\,867\,543^2 = 34\,592\,867\,543 +$ _?_

    A) 0   B) 34592867542 C) 34592867543 D) 34592867544

    31.

32. If 5 oranges cost as much as 2 grapefruits, and 1 grapefruit costs as much as 3 apples, then 10 oranges cost as much as _?_ apples.

    A) 6   B) 9   C) 10   D) 12

    32.

33. The area of the *Winner's Circle* is 4 times that of the other circle, so a radius of the *Winner's Circle* is _?_ as long as a radius of the other.

    A) exactly   B) twice
    C) one-fourth D) 4 times

    33.

34. Of the following, which is largest?

    A) $\sqrt{\frac{1}{2}}$ B) $\sqrt{\frac{1}{4}}$ C) $\frac{1}{4}$ D) $\left(\frac{1}{2}\right)^2$

    34.

35. The second hand of a clock makes _?_ revolutions every 24 hours.

    A) 60   B) 1440   C) 3600   D) 86400

    35.

36. All the following have 2, 3, 5, 6, 10, 15, *and* 30 as factors *except*

    A) 543420   B) 85030   C) 72630   D) 53430

    36.

37. If I save $1 on odd-numbered days every month, and $2 on even-numbered days, how much will I save in the year 2000?

    A) $16   B) $538   C) $545   D) $549

    37.

38. The ones' digit of the largest nine-digit perfect square is

    A) 0   B) 1   C) 4   D) 9

    38.

39. Each km of a 5 km race, my horse's average speed decreased 1 km/hr. If I averaged 5 km/hr at first, it took me _?_ minutes to finish.

    A) 120   B) 137   C) 216   D) 685

    39.

40. I make a straight cut through a rectangular piece of paper and get two pieces. When I make a straight cut through one of these two pieces, I then have three pieces. Altogether, the number of edges on these three pieces *cannot* be

    A) 8   B) 9   C) 10   D) 11

    40.

The end of the contest ✍ **7**

Solutions on Page 85 • Answers on Page 141

2000-2001 Annual 7th Grade Contest

February 20 or 27, 2001

Instructions

7

- **Time** You will have only *30 minutes* working time for this contest. You might be *unable* to finish all 40 questions in the time allowed.

- **Scores** Please remember that *this is a contest, not a test*—and there is no "passing" or "failing" score. Few students score as high as 30 points (75% correct). Students with half that, 15 points, *should be commended!*

- **Format and Point Value** This is a multiple-choice contest. Each answer is an A, B, C, or D. Write each answer in the *Answers* column to the right of each question. A correct answer is worth 1 point. Unanswered questions get no credit. You **may** use a calculator.

	Answers
1. $44\,444 \times 2 = 80\,000 +$? A) 0 B) 8000 C) 8080 D) 8888	1.
2. Round 89.89 to the nearest *tenth*. A) 89.9 B) 89.99 C) 90 D) 90.0	2.
3. $1000 + 0 \times 0 + 1000 =$ A) 0 B) 1000 C) 2000 D) 1 000 000	3.
4. If the weight of 3 tubas is 3 times the weight of 3 trumpets, then to compute the weight of 1 tuba, multiply the weight of 1 trumpet by A) 1 B) 2 C) 3 D) one-third	4.
5. $8+80+800+8000 = 9+90+900+9000 -$? A) 177 B) 999 C) 1000 D) 1111	5.
6. The greatest common factor of 99 and ? is 3. A) 999 B) 1999 C) 2999 D) 3999	6.
7. *What is the smallest number that makes the sentence below true?* If I multiply together *any* ? whole numbers, each greater than 1, but not necessarily different, the product *always* exceeds 2000. A) 6 B) 7 C) 11 D) 12	7.
8. Twice a certain number equals 24. Half that same number equals A) 3 B) 6 C) 12 D) 48	8.
9. $1 \div \frac{1}{10} = 10 \div$? A) 100 B) 10 C) 1 D) $\frac{1}{100}$	9.
10. The length of a rectangle is 2 more than its width. If the rectangle's perimeter is 24, its area is A) 27 B) 35 C) 42 D) 48	10.
11. $3^2 + 6^2 = 9^2 -$? A) 6^2 B) 3^2 C) 2^2 D) 0^2	11.
12. Which of the following has a value different from the other three? A) $2 \times (2^2)$ B) $(2 \times 2)^2$ C) $(2 + 2)^2$ D) $2^2 \times 2^2$	12.
13. The sum of the digits of the product $40 \times 50 \times 60 \times 70$ is divisible by A) 4 B) 5 C) 7 D) 11	13.
14. Which of the following is less than 250%? A) $\sqrt{6.25}$ B) $\frac{5}{2}$ C) 2.5 D) $(0.5)^2$	14.
15. What fractional part of 1 day is 1440 seconds? A) $\frac{1}{12}$ B) $\frac{1}{24}$ C) $\frac{1}{60}$ D) $\frac{1}{3600}$	15.

Go on to the next page ⫸ **7**

16. Of the following, which has the greatest value?
A) $3 \times 2 + 1$ B) $3 + 1 \times 2$ C) $1 \times 3 \times 2$ D) $2 + 3 + 1$ 16.

17. A temperature rise from $-6°$ to $+19°$ is a rise of _?_ degrees.
A) -13 B) 13 C) 25 D) 27 17.

18. 3 cm × 3 cm = (3 × 3 cm) × _?_
A) 1 cm B) 1 cm^2 C) 1 D) 0 cm 18.

19. 1 ten + 5 ones + 15 tenths =
A) 15.15 B) 15.5 C) 16.15 D) 16.5 19.

20. Pat popped 5 fewer Valentine balloons than Pete popped. If Pat and Pete popped a total of 31 balloons, then Pat popped _?_ balloons.
A) 5 B) 13 C) 18 D) 26 20.

21. The reciprocal of $\frac{4}{3}$ has the same value as
A) $\frac{8}{12}$ B) $\frac{12}{16}$ C) $\frac{16}{12}$ D) $\frac{24}{16}$ 21.

22. Which of the following is *not* equal in value to the other three?
A) 200 dimes B) 400 nickels C) 2000 pennies D) 100 quarters 22.

23. $\sqrt{49} + \sqrt{81} =$ A) 2^2 B) 4^2 C) 8^2 D) 16^2 23.

24. What is the sum of 8 numbers whose average is 13?
A) 13 B) 21 C) 64 D) 104 24.

25. 70 × 35¢ is equal to each of the following *except*
A) 2 × 35¢ × 35 B) 0.35 × 70¢ C) 7 × \$3.50 D) \$7 × 3.5 25.

26. The number _?_ is 0.5 more than an odd prime number.
A) 3.5 B) 2.5 C) 1.5 D) 0.5 26.

27. Whenever February has 5 Sundays, the 14th of February will always fall on a
A) Thursday B) Friday
C) Saturday D) Sunday 27.

28. If the sum of 5 whole numbers is 25, then the sum of their square roots is always
A) less than 5 B) equal to 5 C) at least 5 D) at most 5 28.

29. $\frac{2}{5}$ = _?_ % of $\frac{5}{2}$ A) 4 B) 16 C) 20 D) 40 29.

30. Which of the following sums equals a prime number?
A) $12^2 + 33^2$ B) $13^2 + 38^2$ C) $14^2 + 18^2$ D) $15^2 + 49^2$ 30.

Go on to the next page ➡ **7**

31. Rock stars Ewe and Eye were both born on Sunday, January 1. When Ewe was 13 years old, Eye could *not* have been _?_ years old. A) 6 B) 7 C) 8 D) 13	31.
32. If the sum of 10 consecutive whole numbers is divided by 10, the remainder is always A) 0 B) 1 C) 5 D) 9	32.
33. $\sqrt{1 \times 2 \times 3 \times 4 \times 5 \times 6 \times 7 \times 8 \times 9 \times 10}$ = A) $30 + \sqrt{7}$ B) $30 \times \sqrt{7}$ C) $720 + \sqrt{7}$ D) $720 \times \sqrt{7}$	33.
34. If the perimeter of a triangle is 18 and the length of each side is a whole number, then the greatest possible length of one side is A) 7 B) 8 C) 9 D) 10	34.
35. It takes Pat 4 minutes more to run 5 km than it takes Lee to run 3 km. If Lee runs at the rate of $\frac{3}{4}$ km/min, how fast does Pat run? A) $\frac{5}{8}$ km/min B) $\frac{4}{3}$ km/min C) $\frac{5}{4}$ km/min D) 2 km/min	35.
36. How much more than 2^{2000} is 2^{2001}? A) 2 B) 4 C) 2000 D) 2^{2000}	36.
37. $\dfrac{?}{2 + \dfrac{3}{4+5}} = 1 + \dfrac{2}{3+4}$ A) 1 B) 2 C) 3 D) 4	37.
38. A square of side-length 4 is folded in half and then folded in half again. Which could *not* be the length of a side of the new figure? A) 4 B) 3 C) 2 D) 1	38.
39. What is the measure of the smaller angle formed by the hour and minute hands of an accurate circular clock at 12:12 P.M.? A) 48° B) 60° C) 66° D) 72°	39.
40. The mirror images of the calculator digits 1, 2, 5, and 8 are 1, 5, 2, and 8 respectively. For example, the mirror image of 125 is 251. Some number containing the four digits 1, 2, 5, and 8 is added to its mirror image, and their sum is 5000. What is their difference? A) 566 B) 630 C) 1566 D) 1866	40.

The end of the contest ☞ **7**

Solutions on Page 89 • Answers on Page 142

8th Grade Contests

1996-1997 through 2000-2001

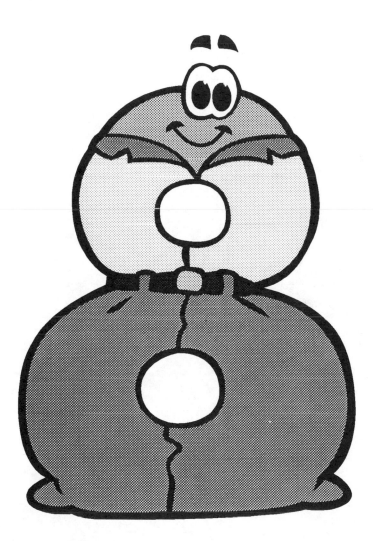

1996-97 Annual 8th Grade Contest

Tuesday, February 4, 1997

Instructions

- **Time** You will have only *30 minutes* working time for this contest. You might be *unable* to finish all 40 questions in the time allowed.

- **Scores** Please remember that *this is a contest, not a test*—and there is no "passing" or "failing" score. Few students score as high as 30 points (75% correct). Students with half that, 15 points, *should be commended!*

- **Format and Point Value** This is a multiple-choice contest. Each answer is an A, B, C, or D. Write each answer in the *Answers* column to the right of each question. A correct answer is worth 1 point. Unanswered questions get no credit. You **may** use a calculator.

1. A positive number minus a negative number is always A) even B) odd C) positive D) zero	1.
2. $3 \div \left(\frac{1}{7} + \frac{2}{7} + \frac{3}{7} + \frac{4}{7} + \frac{5}{7} + \frac{6}{7} \right) =$ A) 1 B) $\frac{3}{7}$ C) 3 D) $\frac{7}{3}$	2.
3. Fifty of Pat's paper airplanes landed in the wastebasket. If Lee was only 80% as effective with wads of paper, how many of Lee's wads of paper landed in the wastebasket? A) 32 B) 40 C) 50 D) 60	3.
4. Of the following numbers, which is the greatest? A) 0 B) –0.5 C) –1 D) –10%	4.
5. How many integers have reciprocals that are also integers? A) 0 B) 1 C) 2 D) 3	5.
6. If it will be 10 A.M. in 11 hours, then it was 10 A.M. _?_ hours ago. A) 10 B) 11 C) 13 D) 24	6.
7. Each of the following is equal to 46 ÷ 6 *except* A) $4\frac{1}{6}$ B) $7\frac{2}{3}$ C) $7\frac{4}{6}$ D) $\frac{46}{6}$	7.
8. The sum 0.123456789 + 0.987654321 contains _?_ nonzero digits. A) 1 B) 7 C) 8 D) 9	8.
9. (the least positive integer) − (the largest negative integer) = A) 2 B) 1 C) 0 D) –1	9.
10. (2112 ÷ 11) − (2002 ÷ 11) = _?_ ÷ 11. A) 2110 B) 110 C) 11 D) 10	10.
11. If a box of 12 candles costs \$4.20, and a box of 6 candles costs \$2.40, what is the least Mom had to pay for the 42 candles she put on the birthday cake she baked for Dad? A) \$14.70 B) \$15 C) \$15.60 D) \$16.80	11.
12. Five-hundredths is _?_ % of one-tenth. A) 2 B) 5 C) 20 D) 50	12.
13. $-3^2 =$ A) 6 B) –6 C) 9 D) –9	13.
14. To the nearest thousandth, 1997 ÷ 1996 = A) 1.01 B) 1.001 C) 1.005 D) 1.0005	14.
15. If a pentagon has p sides and a trapezoid has t sides, then $p \times t =$ A) 15 B) 18 C) 20 D) 24	15.

Go on to the next page ▸ **8**

16. The greatest common factor of 2×4×6×8×10 and 1×2×3×4×5 is A) 2 B) 4 C) 8 D) 120	16.
17. The number 2.5×10^9 is equal to each of the following *except* A) 25×10^8 B) 250×10^8 C) 0.25×10^{10} D) 0.025×10^{11}	17.
18. If Jan is 3 years older now than Dale was 2 years ago, then Jan is _?_ older than Dale is right now. A) 1 year B) 2 years C) 3 years D) 5 years	18.
19. $\frac{21}{1000} + \frac{3}{100} + \frac{4}{10} =$ A) 0.28 B) 0.2134 C) 0.4321 D) 0.451	19.
20. If my cup of tea, with 6% tax added, costs $2.65, then before the tax that cup of tea costs A) $1.65 B) $2.25 C) $2.50 D) $2.53	20.
21. $2^2 + 2^2 \times 2^2 + 2^2 \times 2^2 + 2^2 =$ A) 40 B) 64 C) 84 D) 148	21.
22. If one of the angles of an isosceles triangle has a measure of 50°, what is the measure of the smallest angle of the triangle? A) 25° B) 40° C) 50° D) 80°	22.
23. If I jog 30 minutes every day at 5 km/hr, then I jog _?_ each week. A) 2.5 km B) 17.5 km C) 35 km D) 150 km	23.
24. If the length of each side of a triangle is a multiple of 3, the average length of a side *must* be _?_ number. A) an odd B) an even C) a whole D) a prime	24.
25. Whenever the first and last day of a month fall on the same day of the week, that month has exactly _?_ days. A) 28 B) 29 C) 30 D) 31	25.
26. The number 0.75 is the additive inverse of A) $-\frac{3}{4}$ B) $\frac{3}{4}$ C) $-\frac{4}{3}$ D) $\frac{4}{3}$	26.
27. A couch weighs one-third as much as a couch potato. If their total combined weight is 720 kg, how much does the couch potato weigh? A) 540 kg B) 480 kg C) 240 kg D) 180 kg	27.
28. When a number is _?_ , it is always larger than its reciprocal. A) positive B) negative C) less than 1 D) more than 1	28.
29. In every _?_ triangle, the sum of the measures of two of the angles is equal to the measure of the third angle. A) acute B) obtuse C) isosceles D) right	29.

Go on to the next page ⅢⅢ➡ **8**

30. If the number of spots on a dalmatian equaled the square root of the number of spots on a leopard, a leopard couldn't have _?_ spots.

    A) 121   B) 144   C) 164   D) 169

    30.

31. A certain square has an area of 36. What is the area of a circle whose diameter is as long as the diagonal of that square?

    A) 72π B) 36π C) 24π D) 18π

    31.

32. If I have as many pennies as nickels, then the total value of all these coins could be

    A) $0.50   B) $1.00   C) $1.50   D) $1.75

    32.

33. If the cost of a Funkmobile increases 10% each year, then the cost of a Funkmobile 2 years from now will be _?_% of its current cost.

    A) 120   B) 121   C) 130   D) 230

    33.

34. In a certain sequence, every term after the first two terms is the sum of the two terms that precede it. If this sequence begins 0.1, 0.1, 0.2, 0.3, what is the next term?

    A) 0.3   B) 0.4   C) 0.5   D) 0.6

    34.

35. The length of each side of a right triangle is a whole number of cm. If the hypotenuse is 26 cm long, the shortest side is _?_ long.

    A) 10 cm   B) 12 cm   C) 13 cm   D) 24 cm

    35.

36. $\dfrac{1}{1997} + \dfrac{2}{1997} + \dfrac{3}{1997} + \ldots + \dfrac{1994}{1997} + \dfrac{1995}{1997} + \dfrac{1996}{1997} =$

    A) 998   B) 998.5   C) 1996   D) 1997

    36.

37. The product of the first 25 positive integers ends with _?_ zeros.

    A) 3   B) 4   C) 5   D) 6

    37.

38. The number of worms that Early Bird ate was equal to the number of whole numbers that are divisors of 4^{10}. How many worms did Early Bird eat?

    A) 10   B) 11   C) 20   D) 21

    38.

39. $100-99+98-97+\ldots+4-3+2-1 =$

    A) 49   B) 50   C) 99   D) 100

    39.

40. If the digit-counting starts at page 1, then the total number of digits used to number the pages of my math book could be

    A) 1997   B) 1998   C) 1999   D) 2000

    40.

The end of the contest ✍ **8**

Solutions on Page 95 • Answers on Page 143

1997-98 Annual 8th Grade Contest

Tuesday, February 3, 1998

Instructions

- **Time** You will have only *30 minutes* working time for this contest. You might be *unable* to finish all 40 questions in the time allowed.

- **Scores** Please remember that *this is a contest, not a test*—and there is no "passing" or "failing" score. Few students score as high as 30 points (75% correct). Students with half that, 15 points, *should be commended!*

- **Format and Point Value** This is a multiple-choice contest. Each answer is an A, B, C, or D. Write each answer in the *Answers* column to the right of each question. A correct answer is worth 1 point. Unanswered questions get no credit. You **may** use a calculator.

Question	Answers
1. $(250 + 250 + 250) + (750 + 750 + 750) = 750 \times$? A) 2 B) 3 C) 4 D) 6	1.
2. 6 quarters + 3 dimes = 15 pennies + ? nickels A) 25 B) 30 C) 33 D) 35	2.
3. $100 \times 111 + 10 \times 111 + 1 \times 111 =$? $\times 111$ A) 10 B) 11 C) 110 D) 111	3.
4. Of the following, which is greater than $\frac{5}{16}$? A) 0.4 B) 0.3 C) 0.25 D) 0.2	4.
5. $96 - 48 \div 2 \times 4 =$ A) 96 B) 90 C) 6 D) 0	5.
6. $1 \times 100 \times 2 \times 50 \times 4 \times 25 \times 5 \times 20 =$ A) 400 B) 3×100 C) 100×100 D) 100^4	6.
7. A length of 1 m is ? % of a length of 1 km. A) 0.5 B) 0.1 C) 0.01 D) 0.001	7.
8. What is the volume of 1/3 of 3/4 of a 2ℓ bottle of soda? A) 400 ml B) 500 ml C) 667 ml D) 1500 ml	8.
9. $1122 + 2233 + 3344 =$? $\times (102 + 203 + 304)$ A) 9 B) 10 C) 11 D) 12	9.
10. Huck is twice as old as Becky and 2 years older than Tom. If Becky is 12, how old is Tom? A) 14 B) 20 C) 22 D) 24	10.
11. $\frac{3}{4} =$? $\times \frac{4}{3}$ A) $\frac{9}{16}$ B) $\frac{3}{4}$ C) $\frac{1}{3}$ D) -1	11.
12. 7 days + 7 hours + 7 minutes + 7 seconds = ? seconds A) 10 507 B) 605 647 C) 630 427 D) 630 440	12.
13. What is the ratio of minutes to seconds in one hour? A) 60:1 B) 1:60 C) 1:360 D) 1:3600	13.
14. $100 \div (\frac{1}{10} \times \frac{1}{10}) =$ A) 100×100 B) 10×100 C) 10 D) 1	14.
15. The measure of the smallest angle in an isosceles right triangle is A) 30° B) 45° C) 60° D) 90°	15.
16. The square of a 2-digit integer has at most ? digits. A) 4 B) 3 C) 2 D) 1	16.

Go on to the next page ➡ **8**

17. The difference between two primes, each greater than 100, is
    A) even   B) odd   C) composite   D) prime

17.

18. I make fruit punch with one-third of a cup of each of five different juices. Half this punch fills _?_ cup.
    A) $\frac{1}{5}$ B) $\frac{1}{3}$ C) $\frac{5}{6}$ D) 1

18.

19. How many different positive primes are factors of 1998?
    A) 2   B) 3   C) 4   D) 5

19.

20. Divide the number of cm^2 in the area of a square of side 5 cm by the number of cm in its perimeter. The quotient is
    A) 1.25   B) 1.2   C) 1   D) 0.8

20.

21. If $\overline{BD}$ is the diagonal of square $ABCD$, as shown at the right, then $m\angle ABD =$
    A) 30°   B) 45°   C) 60°   D) 90°

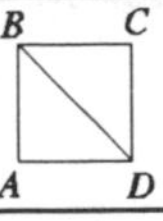

21.

22. What is the product of any number and twice its reciprocal?
    A) 0.5   B) 1   C) 1.5   D) 2

22.

23. Sid's snake slithers 1 km in 50 minutes. Its average speed is
    A) 10 m/min   B) 20 m/min
    C) 30 m/min   D) 50 m/min

23.

24. If $19:98 = x:97$, then
    A) $x < 19$ B) $x > 19$ C) $x = 18$ D) $x = 19$

24.

25. Two lines intersect as shown at the right. If $m\angle 4$ is one-third $m\angle 1$, then $m\angle 3 =$
    A) 135°   B) 120°   C) 60°   D) 45°

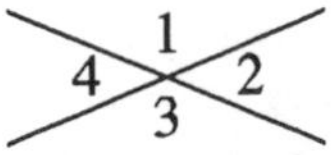

25.

26. If the 14th day of a month is a Wednesday, then the first day of the next month could *not* fall on a
    A) Friday   B) Saturday   C) Sunday   D) Monday

26.

27. $1 \div \frac{1}{2} + 2 \div \frac{1}{4} + 4 \div \frac{1}{8} =$
    A) 1.5   B) 6   C) 8   D) 42

27.

28. The Ivy Club, which reserves 30% of its seats for members, can seat 75 members. How many non-members can it seat?
    A) 150   B) 175   C) 225   D) 250

28.

29. Jane gets 1 magazine every 2 months and another every 6 months. How many issues of these magazines does she receive each year?
    A) 7   B) 8   C) 25   D) 26

29.

30. 21 hours = _?_ % of 1 week.
    A) 87.5   B) 21   C) 12.5   D) 8

30.

Go on to the next page �▶ **8**

33

31. If the sum of 4 consecutive odd numbers is 4 more than the sum of 4 consecutive even numbers, then the least of the odd numbers is _?_ more than the least of the even numbers.

    A) 1   B) 2   C) 3   D) 4   31.

32. If the reciprocal of x is bigger than 1, then

    A) $x < 0$ B) $x = 0$ C) $0 < x < 1$ D) $x > 1$ 32.

The **FAX**...
The Whole **FAX**...
and nothing
but the **FAX**...

33. My cousin Vinnie testified that he was born on a Tuesday in March of a leap year. Long ago, Vinnie told me that his birthday had fallen on a Wednesday exactly twice in his life. What was the youngest age (in years) Vinnie could have been when he told me this?

    A) 6   B) 7   C) 8   D) 9   33.

34. A diameter of one circle is a radius of another. The ratio of their areas, smaller to larger, is

    A) 1:2   B) 1:3   C) 1:4   D) $1 : \pi$ 34.

35. The largest 3-digit integral multiple of 9 equals the largest 3-digit integral multiple of all of the following *except*

    A) 3   B) 17   C) 27   D) 37   35.

36. $\sqrt{1 + \frac{1}{4} + \frac{1}{9}} = 1 + \frac{1}{2} + \frac{1}{3} + \underline{\ ?\ }$ A) $-\frac{2}{3}$ B) 0   C) $\frac{2}{3}$ D) $\frac{7}{6}$ 36.

37. The difference between the area of a square with side-length 2 and the area of any circle drawn inside this square is at least

    A) $4\pi - 4$ B) $4 - 4\pi$ C) $4 - 2\pi$ D) $4 - \pi$ 37.

38. $\frac{1}{2} + \frac{1}{2^2} + \frac{1}{2^3} + \ldots + \frac{1}{2^{18}} + \frac{1}{2^{19}} + \frac{1}{2^{20}} =$ 38.

    A) $\frac{2^{19}}{2^{20}}$ B) $\frac{2^{19} - 1}{2^{20}}$ C) $\frac{2^{19} + 1}{2^{20}}$ D) $\frac{2^{20} - 1}{2^{20}}$

39. I added together all the digits of a certain whole number. If I subtract this sum from my original number, the result of this subtraction *could* be

    A) 1998   B) 1999   C) 2000   D) 2001   39.

40. How many positive numbers less than 10 000 are both squares of integers *and* divisible by 10?

    A) 9   B) 10   C) 99   D) 100   40.

The end of the contest ✍ **8**

Solutions on Page 99 • Answers on Page 144

1998-99 Annual 8th Grade Contest

Tuesday, February 2, 1999

Instructions

- **Time** You will have only *30 minutes* working time for this contest. You might be *unable* to finish all 40 questions in the time allowed.

- **Scores** Please remember that *this is a contest, not a test*—and there is no "passing" or "failing" score. Few students score as high as 30 points (75% correct). Students with half that, 15 points, *should be commended!*

- **Format and Point Value** This is a multiple-choice contest. Each answer is an A, B, C, or D. Write each answer in the *Answers* column to the right of each question. A correct answer is worth 1 point. Unanswered questions get no credit. You **may** use a calculator.

1. $\dfrac{19-99}{99-19}$ = A) –2   B) –1   C) 0.1   D) 1 | 1.

2. $12 \times 20 = 15 \times \underline{?}$
   A) 16   B) 18   C) 20   D) 24 | 2.

3. A gumball machine has 1000 gumballs. At 5¢ per gumball, how much does it cost to empty this machine of its gumballs?
   A) $20   B) $25   C) $50   D) $100 | 3.

4. 500% of $\underline{?}$ is a prime number.
   A) 1   B) 2   C) 3   D) 5 | 4.

5. $(2000 - 1001) + (1000 - 1) = 1999 - \underline{?}$
   A) 1   B) –1   C) 1998   D) 2000 | 5.

6. I squared a nonzero number less than 1. The result must be
   A) less than 1   B) less than 0   C) more than 1   D) more than 0 | 6.

7. $\sqrt{(-1) \times (-9) \times (-9) \times (-9)}$ = A) –27   B) –9   C) 9   D) 27 | 7.

8. In the 20×20 square shown, the shaded region's area is
   A) 10   B) 20   C) 100   D) 200 | 8.

9. $\dfrac{3 \times 9 \times 27 \times 81}{1 \times 3 \times 9 \times 27}$ = A) 3   B) 9   C) 27   D) 81 | 9.

10. If the circle shown is divided into 6 congruent sectors, then each of its 6 central angles is a $\underline{?}$ angle.
    A) 30°   B) 45°   C) 60°   D) 100° | 10.

11. $\sqrt{19 - ?}$ is a whole number.
    A) 3   B) 4   C) 5   D) 6 | 11.

12. For every step Cinderella's godmother takes forward, she takes 2 steps back. If she takes 30 steps forward, she'll wind up $\underline{?}$ steps back from her starting point.
    A) 10   B) 20   C) 30   D) 60 | 12.

13. $4^1 + 4^2 + 4^3 + 4^4 = 4 \times \underline{?}$
    A) 64   B) 85   C) 100   D) 255 | 13.

14. Of the following fractions, which has the smallest reciprocal?
    A) $\dfrac{2}{3}$ B) $\dfrac{4}{3}$ C) $\dfrac{3}{4}$ D) $\dfrac{3}{2}$ | 14.

Go on to the next page ⇒ **8**

15. If the total value of an equal number of pennies, nickels, and dimes is $2.40, how many coins are there altogether?

    A) 60   B) 45   C) 30   D) 15

15.

16. The ratio 12:3 equals the ratio _?_ :8.

    A) 17   B) 24   C) 28   D) 32

16.

17. Working together, my cousin and I can chew through 2 cm of tree trunk every 30 minutes. To chew through a trunk 20 cm thick will take us _?_ hours.

    A) 2.5   B) 5   C) 10   D) 15

17.

18. $\frac{1}{2} \times \frac{1}{4} \times \frac{1}{8} \times \frac{1}{16} = \frac{1}{?}$

    A) 4^2 B) 4^3 C) 4^4 D) 4^5

18.

19. Add the angle-sum of a triangle to the angle-sum of a rhombus.

    A) 360°   B) 540°   C) 720°   D) 900°

19.

20. If the symbol *hh:mm:ss* means *hours:minutes:seconds*, how much time passes from 5:55:55 P.M. until midnight?

    A) 6:04:05   B) 6:05:05   C) 7:04:05   D) 7:05:05

20.

21. $\frac{7}{20} =$ A) 5%   B) 7%   C) 35%   D) 70%

21.

22. Of the following, which is closest to 1?

    A) $0.99 \times 0.01 \times 1.01 \times 0.1$ B) $0.99 \times 0.01 \times 1.01 \times 1$
    C) $0.99 \times 0.01 \times 1.01 \times 10$ D) $0.99 \times 0.01 \times 1.01 \times 100$

22.

23. Multiply the largest 2-digit prime by the smallest 3-digit prime.

    A) 1111   B) 9191   C) 9797   D) 9991

23.

24. If an ant walked 1 km clockwise around a square of side 1 m, how many times did the ant walk around the square?

    A) 1000   B) 250   C) 100   D) 25

24.

25. A circular pizza's 12 slices each have an area of 48π cm². The pizza's radius is

    A) 12 cm   B) 18 cm   C) 24 cm   D) 48 cm

25.

26. 20 000% of 2000% of 200% of 20% of 2 =

    A) 32   B) 320   C) 3200   D) 32 000

26.

27. 2% = _?_ % of 0.1.

    A) 20   B) 2   C) 0.2   D) 0.02

27.

28. _?_ could *not* be the product of 2 of the first 20 whole numbers.

    A) 23   B) 24   C) 30   D) 51

28.

Go on to the next page ▶ **8**

29. Quadrilateral Q's smallest angle is 70°. Its largest angle $\leq$? .
    A) 70°   B) 110°   C) 147°   D) 150°

29.

30. Of the following, which has the largest additive inverse?
    A) $\sqrt{65}$ B) 8   C) 4^2 D) $\sqrt{2}$

30.

31. In a darts tournament, the team whose score was closest to 1 won. The teams' scores are given below. Which team won?
    A) 0.9^{1999} B) 1.1^{1999} C) 1.01^{1999} D) 1.001^{1999}

31.

32. If the area of the right triangle shown is 18, what is the area of the circle?
    A) 36π B) 18π C) 90   D) 36

32.

33. ? cubes with edge 2 will just fit in an empty cube with edge 4.
    A) 2   B) 4   C) 8   D) 64

33.

34. If $a \odot b$ means $a \times (a-b)$, what is the value of $2 \odot 3$?
    A) –2   B) 2   C) 3   D) 6

34.

35. $1\frac{1}{7} \times 1\frac{1}{8} \times 1\frac{1}{9} \times 1\frac{1}{10} \times 1\frac{1}{11} \times 1\frac{1}{12} \times 1\frac{1}{13} =$
    A) 1   B) 2   C) 7   D) 14

35.

36. Robinson Crusoe found a bottle whose only message listed all the prime numbers between 100 and 110. How many numbers were on the message?
    A) 1   B) 2   C) 3   D) 4

36.

37. The number 30 700 is said to end in two 0's. In how many 0's does $2^{1998} \times 5^{1999} \times 10^{2000}$ end?
    A) 2   B) 2000   C) 3998   D) 3999

37.

38. Take some integer and square it. If 15 is a factor of your square, then ? must also be a factor of your square.
    A) 30   B) 75   C) 135   D) 150

38.

39. If the area of each of the nine circles is 4π, what is the area of the square shown?
    A) 144   B) 64   C) 36   D) 36π

39.

40. The area of triangle T is 4. What is the area of a triangle whose sides are each 3 times as long as those of triangle T?
    A) 9   B) 12   C) 36   D) 48

40.

The end of the contest 🖎 **8**

Solutions on Page 103 • Answers on Page 145

1999-2000 Annual 8th Grade Contest

February, 2000

8

Instructions

- **Time** You will have only *30 minutes* working time for this contest. You might be *unable* to finish all 40 questions in the time allowed.

- **Scores** Please remember that *this is a contest, not a test*—and there is no "passing" or "failing" score. Few students score as high as 30 points (75% correct). Students with half that, 15 points, *should be commended!*

- **Format and Point Value** This is a multiple-choice contest. Each answer is an A, B, C, or D. Write each answer in the *Answers* column to the right of each question. A correct answer is worth 1 point. Unanswered questions get no credit. You **may** use a calculator.

1. The average of 1996, 1998, 2000, 2002, and 2004 is A) 5 B) 1999 C) 2000 D) 2500		1.
2. $10 + 110 \times 0 \times 101 + 111 =$ A) 0 B) 120 C) 121 D) 241		2.
3. The difference between any two odd multiples of 3 is always A) even B) odd C) 3 D) 6		3.
4. $\frac{1}{10} \times \left(\frac{1}{10} + \frac{2}{10} + \frac{3}{10} + \frac{4}{10} \right) =$ A) 0.1 B) 10 C) 0.01 D) 0.5		4.
5. Next month, when the cost of an ice cream cone drops $\underline{\ ?\ }$%, its \$2 price will drop to \$1.50. A) 25 B) 33⅓ C) 40 D) 50		5.
6. What is the probability that, in its spelling, a randomly chosen day of the week uses the letter "u"? A) $\frac{2}{7}$ B) $\frac{3}{7}$ C) $\frac{4}{7}$ D) $\frac{5}{7}$		6.
7. Which of the following is greater than 5.25? A) 5.2 B) 52.6% C) $\frac{21}{4}$ D) $\left(\frac{5}{2} \right)^2$		7.
8. When $\frac{392}{5}$ is written in decimal form, its tens' digit is A) 3 B) 4 C) 7 D) 8		8.
9. Of the following, which is nearest in value to 99.900? A) 99.8 B) 99.99 C) 100.000 D) 100.0001		9.
10. 60×60 seconds = A) 36 minutes B) 600 minutes C) 6 hours D) 1 hour		10.
11. Starting with January, number the months from 1 to 12. How many of the odd-numbered months have an even number of days? A) 1 B) 2 C) 3 D) 4		11.
12. My "Wet Paint" sign is a square with a perimeter of 8^2. What is its area? A) 16 B) 32 C) 64 D) 256		12.
13. $\sqrt{\frac{1}{4}} + \sqrt{\frac{1}{4}} + \sqrt{\frac{1}{4}} + \sqrt{\frac{1}{4}} =$ A) 0.5 B) 1.0 C) 1.5 D) 2.0		13.
14. Find the missing number: $\frac{2}{3} \div \frac{3}{2}$ has the same value as $? \times \frac{3}{2}$. A) $\frac{8}{27}$ B) $\frac{2}{3}$ C) $\frac{3}{2}$ D) 1		14.

Go on to the next page ⮕ **8**

15. Ten years ago, Al was as old as Bob is now. In how many years from now will Bob be as old as Al is now? A) 5 B) 10 C) 15 D) 20	15.
16. What is the smallest possible (non-negative) difference between a positive number and its reciprocal? A) 0 B) 1 C) $\frac{1}{4}$ D) $\frac{1}{2}$	16.
17. When Bucky gave me 8 coins worth 25¢ to help him build a dam, he gave me exactly A) 1 nickel B) 2 dimes C) 3 nickels D) 5 pennies	17.
18. Find the missing number: $\frac{100+100}{199+201} = \frac{10}{19+21} + ?$ A) 1 B) $\frac{1}{9+1}$ C) $\frac{100}{19+21}$ D) $\frac{10}{19+21}$	18.
19. If you find the sum of 0.11 and half of 0.11, then half your sum is A) 0.055 B) 0.0825 C) 0.11 D) 0.165	19.
20. A rectangle has area 48 and integral sides. Its perimeter *cannot* be A) 38 B) 42 C) 52 D) 98	20.
21. The sum of 11 different whole numbers can be ? , but never less. A) 11 B) 45 C) 55 D) 66	21.
22. My bicycle wheel is a circle whose area is 400π cm². It's circumference is ? cm. A) 20π B) 40π C) 200π D) 400π	22.
23. 5% of 5% equals 10% of ? . A) 0.025% B) 0.25% C) 2.5% D) 10%	23.
24. A prime number multiplied by its reciprocal will *always* be A) prime B) 0 - C) even D) odd	24.
25. A prime number divided by its reciprocal will *never* be A) prime B) 9 C) even D) odd	25.
26. The perimeter of a triangle is 36 cm. Its longest side *could* be A) 11 cm B) 17 cm C) 18 cm D) 19 cm	26.
27. Cube the negative square root of 4. The result is A) –8 B) 8 C) –6 D) –64	27.
28. I said "no stone shall go unturned" in my search for different whole-number factors of 36. Just how many such factors are there? A) 6 B) 8 C) 9 D) 18	28.

Go on to the next page ⅢⅢ➡ **8**

29. Points A, B, C, and D lie on a line, but not in that order. If $AB = 9$, $BC = 3$, and $CD = 2$, the least possible value of AD is

    A) 2   B) 4   C) 5   D) 7

29.

30. If $a \blacklozenge b$ means $a^2 + b^2$, then $\sqrt{5} \blacklozenge 12 =$

    A) 13   B) 17   C) 60   D) 1691

30.

31. Last night, I kept rockin' for ? hours, a number equal to the least positive difference between two factors of 1 356 724.

    A) 1   B) 2   C) 3   D) 6

31.

32. If $5^{-50} = \dfrac{1}{5^{50}}$, then $1 \div 5^{-50} =$

    A) 5^{50} B) 5^{-50} C) $-\dfrac{1}{5^{50}}$ D) -5^{50}

32.

33. How many numbers equal their reciprocals?

    A) none   B) one   C) two   D) three

33.

34. Three vertices of a certain polygon are A, B, and C. If $AB = AC$, but $AB \neq BC$, then the polygon could be any of the following *except*

    A) a square   B) an isosceles triangle
    C) a rectangle   D) an equilateral triangle

34.

35. Which one of the patterns below *cannot* be folded into a cube in which the three shaded faces all meet at a common vertex?

    A)   B)   C)   D)

35.

36. A semicircle of radius 2 has the same area as a circle of radius

    A) $\dfrac{1}{4}$ B) 1   C) $\sqrt{2}$ D) 2

36.

37. If the average of 2^{1999} and 2^{2001} is equal to the number of pirates who ever lived on Treasure Island, then how many pirates ever lived on Treasure Island?

    A) 5×2^{1998} B) 3×2^{1998} C) 3×2^{1999} D) 2^{2000}

37.

38. If $x^2 < x$, then $\dfrac{1}{x}$ could be equal to

    A) 0.1   B) 0.5   C) 1   D) 2

38.

39. If a *googol* equals 10^{100}, the quotient $1000^{100} \div (1 \text{ googol}) =$

    A) 10 googols   B) 100 googols   C) $(1 \text{ googol})^2$ D) 2 googols

39.

40. For some number n, the sum of the first n positive integers is 240 less than the sum of the first $(n+5)$ positive integers. This number n is itself the sum of the first ? positive integers.

    A) 9   B) 10   C) 45   D) 55

40.

The end of the contest 👈 **8**

Solutions on Page 107 • Answers on Page 146

2000-2001 Annual 8th Grade Contest

February 20 or 27, 2001

Instructions

- **Time** You will have only *30 minutes* working time for this contest. You might be *unable* to finish all 40 questions in the time allowed.

- **Scores** Please remember that *this is a contest, not a test*—and there is no "passing" or "failing" score. Few students score as high as 30 points (75% correct). Students with half that, 15 points, *should be commended!*

- **Format and Point Value** This is a multiple-choice contest. Each answer is an A, B, C, or D. Write each answer in the *Answers* column to the right of each question. A correct answer is worth 1 point. Unanswered questions get no credit. You **may** use a calculator.

1. $10\,100 = 1010 + \underline{\ ?\ }$ A) 9090 B) 9900 C) 9990 D) 19 090	1.
2. $12\,340 = (1 + 2 + 3 + 4) \times \underline{\ ?\ }$ A) 10 B) 100 C) 1010 D) 1234	2.
3. If I get paid $\underline{\ ?\ }$ every 90 seconds, then I'll earn \$100 in an hour. A) 40¢ B) \$1.11 C) \$2.50 D) \$4.00	3.
4. 500 grams $= \underline{\ ?\ } \times 1$ kilograms A) 0.5 B) 2 C) 20 D) 500	4.
5. Of the following, the closest in value to 2.5 is A) $\sqrt{4}$ B) $\sqrt{8}$ C) $(0.5)^2$ D) $\frac{2}{5}$	5.
6. The average of 10, 11, 12, 13, 14, 15, 16, 17, 18, 19, and 20 is A) 15 B) 15.5 C) 16 D) 16.5	6.
7. Add 10 to the number that is 20 less than 5 and you'll get A) –15 B) –5 C) 5 D) 25	7.
8. How many whole numbers less than 10 are squares of themselves? A) one B) two C) three D) four	8.
9. Divide 1 by the sum of $\frac{1}{2}$ and $\frac{3}{4}$, and the quotient will be A) 0.8 B) 1 C) 1.2 D) $\frac{5}{4}$	9.
10. If the sum of the measures of the angles of a certain polygon is 180°, then the average of the measures of these angles is A) 36° B) 45° C) 60° D) 90°	10.
11. Polly sings only on alternate Mondays, so she sings at most $\underline{\ ?\ }$ times in a year. A) 26 B) 27 C) 104 D) 105	11.
12. $(4 \times 2^4) + (2 \times 4^2) = \underline{\ ?\ } \times (2 + 4)$ A) 4^2 B) 4^4 C) 4^6 D) 6^6	12.
13. $(\$8 + 8¢ + \$8.80) \div 8 =$ A) \$2.11 B) \$2.18 C) \$2.20 D) \$3.10	13.
14. Which pair is a pair of *unequal* numbers? A) $\frac{201}{2001}, \frac{67}{667}$ B) $0.0625, \frac{1}{16}$ C) $\frac{105}{30}, \frac{7}{2}$ D) $2.1, \frac{7}{3}$	14.
15. If $\frac{1}{3}$ of my number is 4 more than $\frac{1}{4}$ of my number, my number is A) 24 B) 27 C) 36 D) 48	15.

Go on to the next page ⇒ **8**

16. Sara spent $\frac{1}{3}$ of her cash and lent $\frac{1}{3}$ of the remainder to her sister. What fraction of the original cash does Sara have left?

    A) $\frac{1}{3}$ B) $\frac{2}{3}$ C) $\frac{4}{9}$ D) $\frac{5}{9}$

16.

17. $99 \times 99 \times 99 =$

    A) 99×11^2 B) $27^2 \times 11^3$ C) 9×11^3 D) $3^3 \times 11^3$

17.

18. My school has 24 teachers and 8 grades. If each grade has 30 students, what is the ratio of teachers to students?

    A) 1:10   B) 1:3   C) 4:5   D) 10:24

18.

19. 0.1% is 5% of ? %.

    A) 2   B) 0.02   C) 0.5   D) 0.05

19.

20. $\frac{7+7+7}{14+21+28} = \frac{1}{2} + \frac{1}{3} + \frac{?}{}$

    A) 0   B) $\frac{1}{4}$ C) $\frac{1}{2}$ D) $-\frac{1}{2}$

20.

21. I have 50¢. If I have no pennies, I cannot have ? coins.

    A) 4   B) 7   C) 9   D) 11

21.

22. If two consecutive whole numbers are both prime, their sum is

    A) 1   B) 3   C) 4   D) 5

22.

23. When 2 lines intersect, the sum of the 4 angles thus formed is

    A) 90°   B) 180°   C) 270°   D) 360°

23.

24. The product of a positive number and its additive inverse must be

    A) whole   B) zero   C) positive   D) negative

24.

25. A rope of length 36 fits exactly once around a circle with radius

    A) $\frac{6}{\pi}$ B) $\frac{9}{\pi}$ C) $\frac{18}{\pi}$ D) 4π

25.

26. My mailman said that 30% of last June's rain fell in the first 15 days. What percent of that June's rain fell each day, on average, during the rest of June?

    A) 2%   B) $2\frac{1}{3}$%   C) $3\frac{1}{3}$%   D) $4\frac{2}{3}$%

26.

27. The sum of some numbers is 1234. If one of the numbers is changed from 36 to 63, then the new sum will be

    A) 1198   B) 1207   C) 1261   D) 1270

27.

28. To ride my moped 300 times, I paid $40 for gas. If gas cost $3 per tankful, how many rides per tankful of gas did I average?

    A) 13.3   B) 15   C) 22.5   D) 39

28.

29. What is the least common multiple of 20, 30, 40, 50, and 60?

    A) 300   B) 600   C) 1800   D) 3600

29.

Go on to the next page ⫸ **8**

30. A circle with radius 8 has 4 times the area of a square with side | 30.
    A) $4\sqrt{\pi}$ B) 4π C) 2   D) 16π

31. To color the faces of a cube so that any two faces with an edge in common are colored differently requires at least _?_ different colors. | 31.
    A) 2   B) 3   C) 4   D) 6

32. $(101\,010\,101)^2$ has _?_ non-zero digits. | 32.
    A) 5   B) 6   C) 9   D) 18

33. If the sum of 3 primes is 55, their product could *not* equal | 33.
    A) 705   B) 1334   C) 1505   D) 2001

34. If I and II below are true, which of A), B), C), D) below is also true? | 34.
    I. All 13- and 14-year-old students do well on this contest.
    II. Jan is 11 years old.
    A) Jan might do well.   B) Jan will do well.
    C) Jan will not do well.   D) Jan will not take the contest.

35. The square root of the square root of _?_ is *not* a whole number. | 35.
    A) 0   B) 1   C) 16   D) 100

36. At most how many rectangles with different perimeters can have an area of 216 if the length of every side of every rectangle must be a multiple of 3? | 36.
    A) two B) three C) four D) six

37. If the average of 2 whole numbers is 10, their product *cannot* be | 37.
    A) 0   B) 19   C) 25   D) 91

38. Start at 2001. Count backwards 7 at a time. You can reach | 38.
    A) 1381   B) 1380   C) 1379   D) 1378

39. If C is the midpoint of $\overline{AB}$, D is the midpoint of $\overline{AC}$, and E is the midpoint of $\overline{BC}$, then $\overline{DE}$ has the same length as | 39.
    A) $\overline{AD}$ B) $\overline{BC}$ C) $\overline{CD}$ D) $\overline{CE}$

40. The ones' digit of the *largest* odd factor of $6^{2001} \times 7^{2001} \times 8^{2001}$ is | 40.
    A) 1   B) 3   C) 7   D) 9

The end of the contest ✍ **8**

Solutions on Page 111 • Answers on Page 147

Algebra Course 1 Contests

1996-1997 through 2000-2001

1996-97 Annual Algebra Course 1 Contest

Spring, 1997

Instructions

- **Time** You will have only *30 minutes* working time for this contest. You might be *unable* to finish all 30 questions in the time allowed.

- **Scores** Please remember that *this is a contest, not a test*—and there is no "passing" or "failing" score. Few students score as high as 24 points (80% correct). Students with half that, 12 points, *deserve commendation!*

- **Format and Point Value** This is a multiple-choice contest. Each answer is an A, B, C, or D. Write each answer in the *Answer Column* to the right of each question. A correct answer is worth 1 point. Unanswered questions get no credit. You **may** use a calculator.

1. $(1997)^{1997} + (-1997)^{1997} =$

   A) 0   B) 1997^{3994} C) 2×1997^{1997} D) 3994^{1997}

   1.

2. My n quarters are worth as much as __?__ nickels.

   A) $\frac{n}{5}$ B) $\frac{n}{25}$ C) $5n$ D) $25n$

   2.

3. If $\frac{a}{b} = \frac{1}{2}$, then $\frac{-a}{-b} =$ A) $\frac{-1}{2}$ B) $\frac{1}{-2}$ C) $-\frac{1}{2}$ D) $\frac{1}{2}$

   3.

4. If $x - 1 = 10^3$, then $x^2 - 2x + 1 =$

   A) $10^6 + 1$ B) 10^6 C) $10^9 + 1$ D) 10^9

   4.

5. How many real numbers satisfy $x^2 + 1 = 0$?

   A) none   B) 1   C) 2   D) 8

   5.

6. If $a = \frac{2}{3}$, then $\left(\frac{1}{a}\right)^{-1} = $ __?__.   A) $-\frac{3}{2}$ B) $-\frac{2}{3}$ C) $\frac{2}{3}$ D) $\frac{3}{2}$

   6.

7. $(29x^5 - 23x^4 + 17x^3 - 13x^2 + 11x) - (28x^5 - 24x^4 + 16x^3 - 14x^2 + 10x) =$

   A) $x^5 - x^4 - x^3 - x^2 + x$ B) $x^5 + x^4 + x^3 + x^2 + x$
   C) $x^5 - x^4 + x^3 - x^2 + x$ D) 5

   7.

8. If $x + y = 10$ and $x - y = 8$, then $x^2 - y^2 =$

   A) 2   B) 4   C) 36   D) 80

   8.

9. On our All-Star team, half the number of girls equals one-third the number of boys. The ratio of girls to boys is

   A) 2:3   B) 3:2   C) 3:4   D) 4:3

   9.

10. $x\%$ of $x =$

    A) x^2 B) $\frac{x}{100}$ C) $\frac{x^2}{100}$ D) $100x^2$

    10.

11. $|x - 1| =$

    A) $x - 1$ B) $1 - x$ C) $|x + 1|$ D) $|1 - x|$

    11.

12. The graph of __?__ passes through the point (1996,1997).

    A) $\frac{x}{1996} + \frac{y}{1996} = 2$ B) $\frac{x}{1997} + \frac{y}{1997} = 2$

    C) $\frac{x}{1997} + \frac{y}{1996} = 2$ D) $\frac{x}{1996} + \frac{y}{1997} = 2$

    12.

Go on to the next page ➡ **A**

13. Each of the following is equal to $\sqrt{x^{1997}}$ *except*

    A) $x\sqrt{x^{1995}}$ B) $x^{111}\sqrt{x^{1775}}$ C) $x^{444}\sqrt{x^{1109}}$ D) $x^{789}\sqrt{x^{417}}$

13.

14. If a graph of the results of the Big Bang Research Project is 2 parallel lines, then the product of the lines' slopes *cannot* be

    A) π B) $\sqrt{17}$ C) -1 D) 0

14.

15. The squares of two integers differ by 0. The product of these two integers *could be*

    A) 200   B) 500   C) -800 D) -900

15.

16. Chris and Pat are both teenagers. Today, Chris is c years old and Pat is p years old. If the difference between their ages today is $c - p$, then the difference between their ages 5 years ago was

    A) $c - p$ B) $c - p - 5$ C) $c - (p-5)$ D) $c - p - 10$

16.

17. $\dfrac{(x-1)^{10}}{(1-x)^5} =$ A) $(x-1)^2$ B) $(1-x)^2$ C) $(1-x)^5$ D) $(x-1)^5$

17.

18. $\dfrac{x+y}{\dfrac{1}{x}+\dfrac{1}{y}} =$ A) $x+y$ B) xy C) $\dfrac{1}{xy}$ D) $\dfrac{1}{x+y}$

18.

19. It's 4 P.M. If my raft leaves for Hawaii in h hours, where h is the sum of all integers x which satisfy $4 \le x^2 \le 81$, then my raft leaves for Hawaii at

    A) 4 P.M.   B) 2 P.M.   C) noon   D) 8 A.M.

19.

20. Both roots of $|x| = 2$ are also roots of

    A) $x^2+4x+4 = 0$ B) $x^2-4 = 0$
    C) $x^2-4x+4 = 0$ D) $x^2+4 = 0$

20.

21. What is the sum of the roots of
$(x - 1)(x + 2)(x - 3)\times\ldots\times(x + 98)(x - 99)(x + 100) = 0$?

    A) -50 B) 50   C) -100 D) 100

21.

22. The coordinates of two of the vertices of a square are $(0,0)$ and $(0,-4)$. Of the following, which could *not* be the coordinates of the point where the diagonals of the square intersect?

    A) $(2,-2)$ B) $(-2,2)$ C) $(0,-2)$ D) $(-2,-2)$

22.

Go on to the next page ▐▶ **A**

23. What is the ordered pair of real numbers (U,V) for which
$\dfrac{2}{x^2-4} = \dfrac{U}{x+2} + \dfrac{V}{x-2}$ is true for all $x \neq -2$ or 2?

A) $\left(-\dfrac{1}{2},\dfrac{1}{2}\right)$ B) $\left(-\dfrac{1}{2},-\dfrac{1}{2}\right)$ C) $\left(\dfrac{1}{2},-\dfrac{1}{2}\right)$ D) $\left(\dfrac{1}{2},\dfrac{1}{2}\right)$

23.

24. I measured the perimeter of my square Spring Garden, squared this perimeter, and called my result x. In terms of x, my garden's area is

A) x B) $\dfrac{x}{4}$ C) $\dfrac{x^2}{4}$ D) $\dfrac{x}{16}$

24.

25. If the square of an integer is divisible by 75, it's also divisible by

A) 375   B) 300   C) 225   D) 150

25.

26. If an *equithagorean* triangle is an equilateral triangle in which the sum of the cubes of two sides is equal to the square of the third side, what is the perimeter of an *equithagorean* triangle?

A) 0.5   B) 1.5   C) 4.5   D) 6

26.

27. Of the following, which is true for *all* real numbers a?

A) $\sqrt{a^2} = a$ B) $a^2 \geq a$ C) $a^2+1 \geq 2a$ D) $a \geq \dfrac{1}{a}$

27.

28. What is the area of a circle if the product of its diameter and its radius is 100π?

A) $50\pi^2$ B) 50π C) $100\pi^2$ D) 100π

28.

29. If n is the square of an integer, _?_ is also the square of an integer.

A) $n+1$ B) $n+2\sqrt{n}$ C) n^2+4 D) $n-2\sqrt{n}+1$

29.

30.

1,2,2,3,3,3,4,4,4,4

In the ordered sequence 1, 2, 2, 3, 3, 3, 4, 4, 4, 4, . . . , each positive integer n occurs in a block of n terms. If I add the reciprocals of all the terms, in order beginning with the first term, then at some point the sum will be

A) 98.1   B) 99.25   C) 100.5   D) 102.75

30.

The end of the contest ✍ **A**

1997-98 Annual Algebra Course 1 Contest

Spring, 1998

Instructions

- **Time** You will have only *30 minutes* working time for this contest. You might be *unable* to finish all 30 questions in the time allowed.

- **Scores** Please remember that *this is a contest, not a test*—and there is no "passing" or "failing" score. Few students score as high as 24 points (80% correct). Students with half that, 12 points, *deserve commendation!*

- **Format and Point Value** This is a multiple-choice contest. Each answer is an A, B, C, or D. Write each answer in the *Answer Column* to the right of each question. A correct answer is worth 1 point. Unanswered questions get no credit. You **may** use a calculator.

1. $10x - 9x + 8x - 7x + 6x - 5x + 4x - 3x + 2x - x =$

   A) x B) 5 C) $5x$ D) $10x$

   1.

2. If $x + y = 0$, then $x^2 =$

   A) 0 B) y^2 C) $-y^2$ D) $-x^2$

   2.

3. I'm thinking of a number. Its square is greater than 25, and its cube is less than 125. My number *could* equal

   A) 10 B) 7 C) -5 D) -144

   3.

4. How many values of x satisfy $\dfrac{x^2}{x-2} - \dfrac{4}{x-2} = 0$?

   A) 0 B) 1 C) 2 D) 4

   4.

5. $(x - x)(x^2 - x)(x^3 - x)(x^4 - x)(x^5 - x) =$

   A) $x^{15} - x$ B) $x^{15} - x^5$ C) x^{10} D) 0

   5.

6. $\sqrt{x} + \sqrt{4x} + \sqrt{9x} =$

   A) $\sqrt{6x}$ B) $\sqrt{14x}$ C) $\sqrt{25x}$ D) $\sqrt{36x}$

   6.

7. If $(x + a)(x + 2) = x^2 - x - 6$ for all real numbers x, then $a =$

   A) 3 B) -3 C) 8 D) -8

   7.

8. I played the saxophone for as many hours as the number of real values of x for which $\sqrt{x} + 2 = 0$. I played the saxophone for ? hours.

   A) zero B) one C) two D) four

   8.

9. $x\%$ of $\dfrac{1}{x} =$

   A) x B) 1 C) $\dfrac{1}{x}$ D) $\dfrac{1}{100}$

   9.

10. Multiply the reciprocal of a positive number by the opposite of that same number. The product is

    A) -1 B) 0 C) 1 D) 2

    10.

11. If the lengths of the sides of a quadrilateral are $\sqrt{1}$, $\sqrt{9}$, $\sqrt{9}$, and $\sqrt{8}$, then the perimeter of the quadrilateral is

    A) $\sqrt{1+9+9+8}$ B) $9\sqrt{2}$ C) $7 + 2\sqrt{2}$ D) $18\sqrt{2}$

    11.

Go on to the next page ⇒ **A**

54

12. For what integer x is $17(x+3)$ a positive prime?

A) -2 B) 0 C) 2 D) 1998

12.

13. If a square's perimeter is $36x$, then its area is

A) $24x$ B) $24x^2$ C) $36x^2$ D) $81x^2$

13.

14. $999\,999\,999\,999^2 - 999\,999\,999\,998^2 =$

A) $888\,888\,888\,887$ B) $999\,999\,999\,997$
C) $1\,888\,888\,888\,887$ D) $1\,999\,999\,999\,997$

14.

15. Of the following, which is closest in value to $\sqrt{216}$?

A) $\sqrt{214}$ B) $\sqrt{215}$ C) $\sqrt{217}$ D) $\sqrt{218}$

15.

16. $(x+1)(x+2)(x+3)(x+4) = (x^2+5x+4) \times \underline{\ ?\ }$

A) $x^2 + 5x + 6$ B) $x^2 + 4x + 3$
C) $x^2 + 3x + 2$ D) $x^2 + 2x + 1$

16.

17. Of the following, which *must* be true whenever $x^{1998} = y^{1998}$?

A) $x^{999} = y^{999}$ B) $x^{1997} = y^{1997}$ C) $x^{1999} = y^{1999}$ D) $x^{2000} = y^{2000}$

17.

18. The x-intercepts of $2x - 4y = 16$ and $\underline{\ ?\ }$ are the same.

A) $4x - 2y = 16$ B) $2x + 4y = 16$
C) $4y - 2x = 16$ D) $2y + 4x = 16$

18.

19. The expressions $2n \mid 1$, $2n \mid 3$, and $2n+5$ are *all* positive primes if

A) $n = 0$ only B) $n = 1$ only C) $n = 0$ or 1 D) $n = 2$

19.

20. If the roots of $x^2 + ax + 12 = 0$ are integers, then a is *not*

A) -13 B) -8 C) -7 D) -2

20.

21. One value of x that does *NOT* satisfy $\sqrt{x^2} = x$ is

A) -1 B) 0 C) 1 D) 2

21.

22. What is the area of the right triangle whose vertices have coordinates $(0,1998)$, $(1998,0)$, and $(1998,1998)$?

A) 1998 B) 1998^2
C) 999×1998 D) 0.5×1998

22.

Go on to the next page ⟱ **A**

1997-98 ALGEBRA COURSE 1 CONTEST

23. $x^3 - 1 = (x - 1)(\underline{\ ?\ })$

A) $x^2 - 1$ B) $x^2 + 1$

C) $x^2 + x + 1$ D) $x^2 - x + 1$

23.

24. $\left(a + \dfrac{1}{a}\right)^2 =$

A) $a^2 + \dfrac{1}{a^2}$ B) $a^2 + \dfrac{1}{a^2} + 1$ C) $a^2 + \dfrac{1}{a^2} + 2$ D) $a^2 + \dfrac{1}{a^2} + 4a$

24.

25. If $x \geq 0$ and $y \geq 0$, their *geometric mean* is $\sqrt{xy}$ and their *arithmetic mean* is $\dfrac{x+y}{2}$. The *arithmetic mean* equals the *geometric mean* whenever

A) $x > y$ B) $x = y$ C) $x < y$ D) $x = 0$

25.

26. If the lengths of the sides of a rectangle are integers, and if one side is 2 more than another, then the perimeter *could* equal

A) 1998   B) 1999   C) 2000   D) 2002

26.

27. If $ab = 2$, then $(a - \dfrac{1}{b}) + (b - \dfrac{1}{a}) =$

A) $a + b$ B) $\dfrac{a+b}{2}$ C) $\dfrac{a+b}{3}$ D) $\dfrac{a+b}{4}$

27.

28. If n is a positive integer, $n!$ is the product of the first n positive integers. For example, $4! = 4 \times 3 \times 2 \times 1 = 24$. If u and v are positive integers and $u! = v! \times 56$, then v could equal

A) 6   B) 8   C) 56   D) 57

28.

29. If $-2 < x \leq 1$ and $-1 < y \leq 2$, then it is possible that $xy =$

A) 6   B) 5   C) 4   D) –3

29.

30. I found the product of n different positive prime numbers. Of the following, which could be the number of positive integers that are factors of this product?

A) 1024   B) 1000   C) 676   D) 400

30.

The end of the contest ☞ **A**

Solutions on Page 121 • Answers on Page 149

1998-99 Annual Algebra Course 1 Contest

Spring, 1999

Instructions

- **Time** You will have only *30 minutes* working time for this contest. You might be *unable* to finish all 30 questions in the time allowed.

- **Scores** Please remember that *this is a contest, not a test*—and there is no "passing" or "failing" score. Few students score as high as 24 points (80% correct). Students with half that, 12 points, *deserve commendation!*

- **Format and Point Value** This is a multiple-choice contest. Each answer is an A, B, C, or D. Write each answer in the *Answer Column* to the right of each question. A correct answer is worth 1 point. Unanswered questions get no credit. You **may** use a calculator.

1998-99 ALGEBRA COURSE 1 CONTEST

1. If $x + 1997 = 1998$, then $x + 1999 =$

   A) 1   B) 2000   C) 2001   D) 2002

2. If $x^2 = 16$, the two possible values of $(x - 1)^2$ are

   A) 9 & –25   B) –9 & –25   C) –9 & 25   D) 9 & 25

3. Since they didn't know any algebra, I had to teach my friends at the water fountain that $(1-x)$ and $(1+x)$ have a product equal to

   A) $(x-1)(1+x)$ B) $-(x-1)(x+1)$
   C) $-(1-x)(1+x)$ D) $(x-1)(x+1)$

4. $1999 + 1999 - 1999 = 1999 - \underline{\ ?\ }$

   A) 0   B) 1   C) 2×1999 D) –1999

5. $1 - 2 + 3 - 4 + 5 - 6 + 7 - 8 + 9 - 10 =$

   A) 0   B) –1   C) –5   D) –10

6. Which is 25% less than x?

   A) $0.25x$ B) $25x$ C) $0.75x$ D) $75x$

7. If $m - a + t - h = 0$, then $a =$

   A) $h - m - t$ B) $m + h - t$ C) $m + t - h$ D) $m + h + t$

8. The number $\underline{\ ?\ }$ could be the sum of three consecutive integers.

   A) 0   B) 2   C) 4   D) 8

9. What is the product of the roots of $(x - 3)(x - 4) = 0$?

   A) 7   B) 12   C) –7   D) –12

10. If x and y are reciprocals of each other, then their product always equals

    A) 1   B) x^2 C) y^2 D) $\dfrac{1}{x^2}$

11. The average of 1999 and $\underline{\ ?\ }$ is 1.

    A) –2001   B) –2000   C) –1998   D) –1997

12. Both n and $(\sqrt{n - 1998})(\sqrt{1999 - n})$ are integers when $n =$

    A) 0   B) 1998 only   C) 1999 only   D) both 1998 & 1999

Go on to the next page ⫸ **A**

13. If the **SUM** of 1999 real numbers is –1998, at least how many of these 1999 real numbers *must be* positive?

    A) none   B) 1   C) 1997   D) 1998

13.

14. If the **PRODUCT** of 1999 real numbers is –1998, at least how many of these 1999 real numbers *must be* negative?

    A) none   B) 1   C) 1997   D) 1998

14.

15. In a recent Math Duel, my team began by graphing $51x+24y = 75$. If our opponents graphed a line parallel to ours, they might have graphed

    A) $17x+16y = 25$ B) $17x+8y = 50$
    C) $34x+8y = 50$ D) $34x+16y = 50$

15.

16. If a is a positive number, then $(a + \frac{1}{a})^2 = a^2 + \frac{1}{a^2} + \underline{?}$

    A) a B) 0   C) 1   D) 2

16.

17. If one of five consecutive even integers is $a+3$, another could be

    A) $a - 9$ B) $a - 7$ C) $a - 5$ D) $a - 2$

17.

18. Our spring break lasted only $a+b$ days, where a and b are positive integers and $\sqrt{a} + \sqrt{b} = \sqrt{8}$. Our break lasted $\underline{?}$ days.

    A) 4   B) 6   C) 8   D) 16

18.

19. If $x < 0$, then $|x|$ is equal to

    A) x^2 B) x C) $x-2x$ D) $-2x$

19.

20. If $x < 0$, then $\sqrt{x^2}$ is equal to

    A) $-x$ B) x C) $\sqrt{-x^2}$ D) $-\sqrt{x^2}$

20.

21. If the coordinates of the vertices of a parallelogram are $(-2,3)$, $(3,1)$, $(-1,4)$, and $(-6,6)$, then one of its diagonals has a slope of

    A) -1 B) $\frac{-2}{5}$ C) $\frac{5}{9}$ D) 1

21.

Go on to the next page ▐▶ **A**

59

22. If one leg and the hypotenuse of a right triangle have lengths $\sqrt{a}$ and $\sqrt{c}$, with $a < c$, then the other leg's length is

    A) $\sqrt{c^2 - a^2}$ B) $\sqrt{c^2 + a^2}$ C) $\sqrt{c - a}$ D) $c + a$

22.

23. I proved that exactly _?_ different positive integers n satisfy $14\,400 \le n^2 \le 16\,900$.

    A) 2   B) 9   C) 10   D) 11

23.

24. Whenever $\frac{1999}{x} < 1$, it is *always* true that

    A) $\frac{1}{x} < \frac{1}{1999}$ B) $\frac{x}{1999} > 1$ C) $x > 1999$ D) $\frac{1998}{x} < 0$

24.

25. If the sum of two unequal positive integers is 2000, then the sum of their reciprocals is smallest when the two integers are

    A) 996 & 1004   B) 997 & 1003   C) 998 & 1002   D) 999 & 1001

25.

26. The product 1.25×10^n is the cube of an integer when $n =$

    A) 1999   B) 2000   C) 2001   D) 2002

26.

27. What is the largest power of 5 which divides the product of the first 100 positive integers: $1 \times 2 \times 3 \times \ldots \times 98 \times 99 \times 100$?

    A) 5^2 B) 5^{20} C) 5^{24} D) 5^{100}

27.

28. My classroom has as many temporary repairs as the x-intercept of the graph of $y = \frac{3}{2x-1} - \frac{1}{x-2}$. How many is that?

    A) 7   B) 5   C) 3   D) 2

28.

29. In a certain right triangle, the lengths of the legs have a sum of 50 and a product of 282. What is the length of the hypotenuse?

    A) 44   B) 42   C) 40   D) $25\sqrt{2}$

29.

30. The only real root of $x^3 - 1999x^2 + x - 1999 = 0$ is 1999. What is the only real root of $(x-1)^3 - 1999(x-1)^2 + (x-1) - 1999 = 0$?

    A) 1   B) 1998   C) 1999   D) 2000

30.

The end of the contest ✍ **A**

Solutions on Page 125• Answers on Page 150

1999-2000 Annual Algebra Course 1 Contest

Spring, 2000

Instructions

- **Time** You will have only *30 minutes* working time for this contest. You might be *unable* to finish all 30 questions in the time allowed.

- **Scores** Please remember that *this is a contest, not a test*—and there is no "passing" or "failing" score. Few students score as high as 24 points (80% correct). Students with half that, 12 points, *deserve commendation!*

- **Format and Point Value** This is a multiple-choice contest. Each answer is an A, B, C, or D. Write each answer in the *Answer Column* to the right of each question. A correct answer is worth 1 point. Unanswered questions get no credit. You **may** use a calculator.

1999-2000 ALGEBRA COURSE 1 CONTEST

1. If $x = 2000$, then $(x - 1999)(1999 - x) =$

   A) -1 B) 0 C) 1 D) 2000

   1.

2. The reciprocal of $0.5x$ is

   A) $\dfrac{2}{x}$ B) $2x$ C) $0.5x$ D) $-0.5x$

   2.

3. Dr. Roentgen, who began to study my x-ray at noon, finished $-1-(-1)-(-1)$ hours later, at

   A) 1 P.M.   B) 2 P.M.   C) 3 P.M.   D) 4 P.M.

   3.

4. $(x - 3)(x + 4) = (x + 3)(x - 4) +$?

   A) x B) $2x$ C) $-x$ D) $-2x$

   4.

5. How long is a side of a regular hexagon whose perimeter equals the perimeter of an equilateral triangle whose side is x cm long?

   A) $0.5x$ cm   B) x cm   C) $1.5x$ cm   D) $2x$ cm

   5.

6. If $(a - 1)^2 = (a + 1)^2$, then $a =$

   A) -0.25 B) 0 C) 0.25 D) 0.5

   6.

7. The product of the digits of 2000 is 0. How many integers y, $2000 < y < 2100$, also have 0 as the product of their digits?

   A) 19 B) 20 C) 99 D) 100

   7.

8. $x \div y = y \div$?

   A) $\dfrac{y}{x}$ B) $\dfrac{x}{y}$ C) $\dfrac{y^2}{x}$ D) $\dfrac{x}{y^2}$

   8.

9. I say that my new apartment number is $\sqrt{1 + 3 + 5 + \ldots + 17 + 19}$. Most people say it's

   A) 10 B) 19 C) 20 D) 50

   9.

10. If $x \neq 0$, and $x^2 = -2000x$, then $x =$

    A) -2000 B) -1 C) 1 D) 2000

    10.

11. If $x > 1$, then $\dfrac{1}{x-1} - \dfrac{x}{x-1} = \dfrac{?}{x+1}$ if the missing expression is

    A) $x + 1$ B) $x - 1$ C) $1 - x$ D) $-x - 1$

    11.

12. The product of 2000 negative numbers *could* equal

    A) -2000 B) -1 C) 0 D) π

    12.

Go on to the next page ⯈ **A**

13. $\left(\frac{1}{x} + \frac{1}{x^2}\right) \div (x + x^2) =$

A) $(x+1)^2$ B) $\frac{1}{x^3}$ C) 1   D) 2

13.

14. I hoped the graph would be a line with x-intercept 1 and y-intercept 1. The equation of such a line is

A) $x - y = 1$ B) $x + y = 1$
C) $x + y = -1$ D) $y = x + 1$

14.

15. $\left(\frac{1}{x}\right)\left(\frac{x}{x^2}\right)\left(\frac{x^2}{x^3}\right)\left(\frac{x^3}{x^4}\right)\left(\frac{x^4}{x^5}\right)\left(\frac{x^5}{x^6}\right) =$

A) 1   B) x C) $\frac{1}{x}$ D) $\frac{1}{x^6}$

15.

16. If $3x:5y = 7:11$, then $x:y =$

A) $2:3$ B) $21:55$ C) $35:33$ D) $55:21$

16.

17. $\dfrac{\left(x^{2000}\right)^{2000}}{\left(x^{2000}\right)\left(x^{2000}\right)} =$ A) 1   B) x^{2000} C) x^{4000} D) $x^{3\,996\,000}$

17.

18. How many integers $n < 100$ satisfy $n + 200 > 0$?

A) 100   B) 298   C) 299   D) 300

18.

19. I bought 1 broccoli spear, 1 banana, and 1 orange. The price I paid averaged 93¢ each for the broccoli and the banana, but only 81¢ each for the broccoli and the orange. The banana cost _?_ more than the orange.

A) 4¢   B) 6¢   C) 12¢   D) 24¢

19.

20. $(1 - x^2) \div (x + 1) =$

A) $x + 1$ B) $1 - x$ C) $-x$ D) $x - 1$

20.

21. If a, b, c, d are 4 different numbers, then the line joining $R(-a,b)$ and $S(-c,-d)$ has the same slope as the line joining $T(a,-d)$ and

A) (c,b) B) $(-c,b)$ C) $(c,-b)$ D) $(-c,-b)$

21.

22. $1 - 2 + 3 - 4 + \ldots + 1997 - 1998 + 1999 - 2000 =$

A) -2000 B) -1000 C) -1 D) 0

22.

Go on to the next page ▐▶ **A**

23. For a rectangle with dimensions ℓ and w, the ratio of the numerical value of its perimeter to the numerical value of its area is

   A) $(\ell+w):\ell w$ B) $\ell w:(2\ell+2w)$ C) $\left(\frac{2}{\ell}+\frac{2}{w}\right):1$ D) $2:1$

   23.

24. If a and b are integers, what is the least integer x for which the inequality $a < b < a + x$ could be true?

   A) -1 B) 0 C) 1 D) 2

   24.

25. When $n = \underline{\ ?\ }$, one term in the expansion of $(2x^2+3x+2)^n$ is 1024.

   A) 4 B) 8 C) 9 D) 10

   25.

26. If my rocketcar goes 30 km at an average speed of 30 km/hr, what average speed must it maintain over the next 30 km for it to achieve an average speed of 50 km/hr for the entire 60 km?

   A) 70 km/hr   B) 80 km/hr   C) 100 km/hr   D) 150 km/hr

   26.

27. The roots of $x^2-999999x-9999 = 0$ and $\underline{\ ?\ }$ are additive inverses.

   A) $x^2+999999x-9999 = 0$ B) $x^2-999999x+9999 = 0$
   C) $x^2+999999x+9999 = 0$ D) $x^2-9999x-999999 = 0$

   27.

28. Each of the letters t, h, a, n, k, and s is assigned a different one of the values -1, -2, -3, -4, -5, and -6. What is the least possible value of $t^1+h^2+a^3+n^4+k^5+s^6$?

   A) -209 B) -2229 C) -7879 D) -7897

   28.

29. $(3^{1+x})(3^{1-x}) =$

   A) 3 B) 9 C) 3^{1-x^2} D) 9^{1-x^2}

   29.

30. There are $\underline{\ ?\ }$ different ways to represent $\sqrt{1000000}$ in the form $a\sqrt{b}$, where a and b are positive integers. One way is $1\sqrt{1000000}$.

   A) 2 B) 4 C) 12 D) 16

   30.

The end of the contest ✍ **A**

Solutions on Page 129 • Answers on Page 151

2000-2001 Annual Algebra Course 1 Contest

Spring, 2001

Instructions

- **Time** You will have only *30 minutes* working time for this contest. You might be *unable* to finish all 30 questions in the time allowed.

- **Scores** Please remember that *this is a contest, not a test*—and there is no "passing" or "failing" score. Few students score as high as 24 points (80% correct). Students with half that, 12 points, *deserve commendation!*

- **Format and Point Value** This is a multiple-choice contest. Each answer is an A, B, C, or D. Write each answer in the *Answer Column* to the right of each question. A correct answer is worth 1 point. Unanswered questions get no credit. You **may** use a calculator.

2000-2001 ALGEBRA COURSE 1 CONTEST

1. If $x = 10$, then $(x - 1)^2 + (1 - x)^2 =$

   A) 162   B) 81   C) 9   D) 0

2. If a pushpin is placed at the intersection of the lines $x = -1$ and $y = 2$, it will be at the point

   A) (1,2)   B) (-1,2)   C) (1,-2)   D) (-1,-2)

3. If $x + y = 0$, then $x^2 =$

   A) 0   B) y^2 C) $-y^2$ D) $-x^2$

4. Of the following values of x, the one that makes $\sqrt{1 - x}$ largest is

   A) 1   B) $\frac{1}{4}$ C) $\frac{5}{16}$ D) $\frac{15}{64}$

5. If three vertices of a square have coordinates (2001,1), (4001,1), and (2001,2001), then the coordinates of the fourth vertex are

   A) (1,2001)   B) (2001,4001)   C) (4001,2001)   D) (4001,4001)

6. The product of the roots of $(x - 3)(x - 667) = 0$ is

   A) 770   B) -770   C) 2001   D) -2001

7. How many real numbers x satisfy $\frac{1}{100} < \frac{1}{x} < \frac{1}{99}$?

   A) none   B) 1   C) 2   D) more than 2

8. Please help the *Information Desk* volunteer. She needs to know which of the following *cannot* be a prime number whenever p is a prime number greater than 2001.

   A) $p - 2$ B) $p + 2$

   C) $p + 999$ D) $p + 9990$

9. If $m - a = t - h$, then

   A) $m-t = a-h$ B) $m+t = a-h$ C) $m-t = a+h$ D) $m+t = a+h$

10. If $(x - 1)^2 = 2000^2$ and $(x + 1)^2 = 2002^2$, then $x =$

    A) -2003   B) -2001   C) -1999   D) 2001

11. If $(x + 1)^2 = 2000^2$ and $(x - 1)^2 = 2002^2$, then $x =$

    A) -2003   B) -2001   C) -1999   D) 2001

Go on to the next page ▦➡ **A**

12. If $2000x - 2000 = 1000$, then $2001x - 2001 =$

A) 1000.5   B) 1001   C) 1001.5   D) 1002

12.

13. The graph of $x = -1$ is perpendicular to the graph of

A) $x = 1$ B) $x = 0$ C) $x + y = 1$ D) $y = 2$

13.

14. I ate enough scoops of ice cream to meet ¼ of my daily goal. Had I eaten an additional 2.5 scoops, I would have met ⅓ of my daily goal of _?_ scoops.

A) 12   B) 24   C) 27   D) 30

14.

15. $\sqrt{2000^2 - (2)(2000)(2001) + 2001^2} =$

A) 2000   B) $2001 - 2000$

C) $2000 - 2001$ D) 2001

15.

16. Which of the following is a factor of $1 + (x + 1)(x + 2) + x$?

A) x B) $x + 2$ C) $x + 3$ D) $x + 4$

16.

17. The sum of the roots of $x^2 - 98x - 99 = 0$ is

A) -99 B) -98 C) 98   D) 99

17.

18. How many real values of n satisfy $\sqrt{n} + \sqrt{n} = \sqrt{2n}$?

A) none   B) 1   C) 2   D) more than 2

18.

19. Of the following, which is *not* a factor of $x^4 - 16$?

A) $x^2 + 4$ B) $x^2 - 4$ C) $x - 2$ D) $(x + 2)^2$

19.

20. The two of us ordered a pizza whose area was the same as the area of a square in which the sum of squares of the four sides plus the sum of the squares of the two diagonals is 400. The pizza's area was

A) 50   B) 100   C) 150   D) 200

20.

21. $x^{2000} + (-x)^{2000} =$

A) $2x^{2000}$ B) $2x^{2001}$ C) $|x|^{2001}$ D) 0

21.

22. The least common multiple of x^2, x^3, and x^4 is

A) x^4 B) x^6 C) x^{12} D) x^{24}

22.

Go on to the next page ⫸ **A**

23. If $|a| = -a$ and $|b| = -b$, then $a - b =$

    A) $|a| + |b|$ B) $|a| - |b|$ C) $-|a| + |b|$ D) $-|a| - |b|$

    23.

24. Which of the following is an equation of a line whose slope is twice the slope of the line whose equation is $2x - y = 3$?

    A) $4x - y = 3$ B) $2x - 2y = 3$
    C) $4x - 2y = 6$ D) $2x - 2y = 6$

    24.

25. How many integers greater than 1 million and less than 4 million are squares of integers?

    A) 998   B) 999   C) 1000   D) 1998

    25.

26. For how many different negative values of x is $\sqrt{x + 2001}$ an integer?

    A) 42   B) 43   C) 44   D) 45

    26.

27. If the value of $21x + 28y = 84$, then the value of $6x + 8y$ is

    A) 12   B) 20   C) 24   D) 48

    27.

28. In a *pseudopythagorean* triangle, the sides are all integers and the sum of the square roots of two of the sides equals the square root of the third side. How many *pseudopythagorean* triangles are there?

    A) none   B) 1   C) 2   D) more than 2

    28.

29. If $5^x = 7$, then $5^{x+2} =$

    A) 9   B) 14   C) 49   D) 175

    29.

30. In how many of the ordered pairs of positive integers (x,y) that satisfy $2x + 3y = 9\,999\,999\,999\,999$ is x a prime number?

    A) none   B) one   C) two   D) three

    30.

The end of the contest ✐ **A**

Solutions on Page 133 • Answers on Page 152

Detailed Solutions

●●●●●●●●●●●●●●●●●●

1996-1997 through 2000-2001

7th Grade Solutions

1996-1997 through 2000-2001

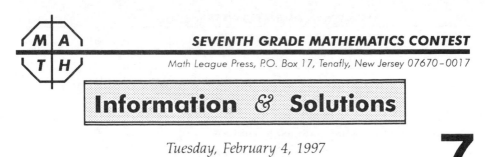
Information & Solutions

Tuesday, February 4, 1997

7

Contest Information

- **Solutions** Turn the page for detailed contest solutions (written in the question boxes) and letter answers (written in the *Answers* column to the right of each question).

- **Scores** Please remember that *this is a contest, not a test*—and there is no "passing" or "failing" score. Few students score as high as 30 points (75% correct). Students with half that, 15 points, *deserve commendation!*

- **Answers & Rating Scale** Turn to page 138 for the letter answers to each question and the rating scale for this contest.

1. $10 \times (73 + 37) = 10 \times 110 = 10 \times (91 + 19)$, so the answer is B. A) 9 B) 19 C) 37 D) 110	1. B
2. Convert to 64ths by hand **or** convert to a decimal by calculator. A) $\frac{3}{4} = \frac{48}{64}$ B) $\frac{5}{8} = \frac{40}{64}$ C) $\frac{11}{16} = \frac{44}{64}$ D) $\frac{37}{64}$	2. A
3. Add the digits. If their sum is divisible by 9, so is the number. A) 18 018 B) 18 081 C) 81 180 D) 81 181	3. D
4. Seven months have 31 days. By calculator, $7/12 = 0.58333 \ldots$. A) 75% B) 70% C) 58% D) 50%	4. C
5. The first 72 singers were split into groups of 4. There were 18 such groups. The 73rd singer starts group 19. A) 18 B) 19 C) 20 D) 73	5. B
6. Note that 25% of 8 is 2, and 2 is 50% of 4. A) 64 B) 16 C) 4 D) 2	6. C
7. The paper forms a right triangle when folded in half diagonally. Unfold it. You'll get two right triangles that fill a square. A) square B) triangle C) trapezoid D) pentagon	7. A
8. \$2.20 with 10 coins? The average coin is worth 22¢. Use 8 quarters. A) pennies B) nickels C) dimes D) quarters	8. D
9. 60×60 sec. in an hr., but 60 sec. in a min. Finally, $(60 \times 60)/60 = 60$. A) 24 B) 60 C) 360 D) 3600	9. B
10. A = 10010, B = 1010, C = 101, and D = 110. The smallest is C. A) 1.001×10^4 B) 1.01×10^3 C) 1.01×10^2 D) 1.1×10^2	10. C
11. The product of reciprocals is 1, and a 0 can't be in that product. A) 0 B) 0.5 C) 1 D) 2	11. A
12. 404.404 rounds to 404.4, and we don't want a hundredth's digit. A) 400 B) 404.5 C) 404.40 D) 404.4	12. D
13. 404.404 rounds to 404.40, and we must have a hundredth's digit. A) 400 B) 404.5 C) 404.40 D) 404.4	13. C
14. $11/18 = 0.6111 \ldots$. It's nearest to 0.62. A) $\frac{6}{10}$ B) $\frac{7}{11}$ C) $\frac{10}{17}$ D) $\frac{11}{18}$	14. D
15. Add 5 to 21. That sum is twice Pat's age today. Divide the sum, 26, by 2 to get 13. That's Pat's age today. A) 12 B) 13 C) 14 D) 15	15. B

Go on to the next page ⫸ **7**

16. When both primes are odd, their sum is even; so one prime is a 2. A) 5 B) 3 C) 2 D) 1 (not prime)	16. C
17. In all the choices, *two* of the numbers have a product of 24. But only in choice A is the sum of all *three* numbers equal to 24. A) 3, 8, 13 B) 4, 6, 16 C) 2, 12, 12 D) 1, 24, 1	17. A
18. The sum of all 3 angles of a triangle is 180°, so their average is 60°. A) 45° B) 60° C) 90° D) 180°	18. B
19. 30¢ ÷ \$5 = 30¢ ÷ 500¢ = 30/500 = 3/50 = 6/100 = 6%. A) 5% B) 6% C) 8% D) 30%	19. B
20. If we subtract 281 from both page numbers, it's clear that Volume Two goes from page 1 to page 281, a total of 281 pages. A) 279 B) 280 C) 281 D) 282	20. C
21. To see which is different, factor each into primes. Each equals $3^2 \times 5 \times 7$ except choice A. (Or compute each with a calculator.) A) 13×25=325 B) 7×45=315 C) 15×21=315 D) 9×35=315	21. A
22. To celebrate the first day of a leap year, I taught my dog to jump through a hoop. It was a Sunday. 366 days later is 52 weeks + 2 days later; and 2 days after Sunday is Tuesday. A) Sunday B) Monday C) Tuesday D) Wednesday	22. C
23. As in *dec*imal, "*dec*" means "ten;" so a *deca*gon has 10 sides. A) 5 B) 6 C) 8 D) 10	23. D
24. It's just like a scale drawing. All lengths are being scaled down by a factor of 2. If a string forms a circle with an 8 cm diameter, then half the string will form a circle with a 4 cm diameter. A) 4/π cm B) 8/π cm C) 2 cm D) 4 cm	24. D
25. By calculator, $\frac{3}{4} \times \frac{7}{8} \times \frac{15}{16}$ = 0.61 . . . , which is nearest to 0.50. A) 0.25 B) 0.50 C) 0.75 D) 1.00	25. B
26. I drank cola on Dec. 3-6, 9-13, 16-20, 23-27 for a total of 19 glasses. I had my next 3 colas on Dec. 30, 31, and Jan. 1. A) Dec. 24 B) Dec. 25 C) Dec. 31 D) Jan. 1	26. D
27. 25% of 100×100 = (100×100)/4 = (100/2)(100/2) = 50×50. A) 25×25 B) 50×50 C) 200×200 D) 400×400	27. B
28. Since 5×25 = 125, the product of 5 and its square, 25, is 125. A) 5 B) 10 C) 25 D) $\sqrt{125}$	28. A
29. $6^8 \times 10^6 = (2\times3)^8 \times (2\times5)^6 = 2^8 \times 3^8 \times 2^6 \times 5^6$. Now remove all the 2's. A) $(6^8 \times 10^6)/2$ B) $3^4 \times 5^3$ C) 15^6 D) $3^8 \times 5^6$	29. D

Go on to the next page ▐▐▐▶ **7**

30.	The hypotenuses of four congruent isosceles right triangles are the sides of a square. The 4 shaded regions can be matched up to the 4 unshaded regions. Their areas must be equal. A) 36 B) 27 C) 18 D) 9	30. C
31.	December has 31 days, November has 30 days, October has 31 days, and September has 30 days. All together, that's 122 days. Going backwards, the next day is in August. A) August B) July C) March D) September	31. A
32.	Each wheel's circumference is 120π, so number of revolutions = (total distance)/(circumference) = $(500 \text{ m})/(120\pi \text{ cm}) = (50\,000 \text{ cm})/(120\pi \text{ cm})$. A) $\frac{500}{120\pi}$ B) $\frac{1000}{120\pi}$ C) $\frac{50\,000}{120\pi}$ D) $\frac{100\,000}{120\pi}$	32. C
33.	Any 5 such numbers add up to a multiple of 5. A) 225 B) 222 C) 220 D) 200	33. B
34.	There are 6 ways with Di standing first: DBJM, DBMJ, DJBM, DJMB, DMBJ, & DMJB. There are 6 ways with *anyone* standing first. A) 4 B) 6 C) 16 D) 24	34. D
35.	$10^{18}/2 = (2^{18}\times5^{18})/2 = 2^{17}\times5^{18} = 5\times2^{17}\times5^{17} = 5\times10^{17}$, or use a calculator. A) 5^9 B) 5^{18} C) 10^9 D) 5×10^{17}	35. D
36.	Since Bob is lying, Ann is lying and Carol is telling the truth. Since Carol is telling the truth, Dan is telling the truth. Of the three other than Bob, only Ann is lying. A) 0 B) 1 C) 2 D) 3	36. B
37.	If the hands move *counterclockwise*, then 20 minutes before the hour hand points to the 3, the minute hand points to 20 minutes beyond 12, which is 4. A) 4 B) 7 C) 8 D) 11	37. A
38.	The ones' digits cycle 2,4,8,6,2,4,8,6 .... The 53rd digit is a 2. A) 2^{53} B) 2^{52} C) 2^{51} D) 2^{50}	38. A
39.	Unfolding all 5 of the tape's folds doubles its length 5 times, so the length of the tape is 5 cm $\times2\times2\times2\times2\times2 = 160$ cm. A) 25 cm B) 80 cm C) 100 cm D) 160 cm	39. D
40.	Pat paid 1/8 and Jan paid 1/4, so they paid 3/8 of the bill. I paid 5/8 of the bill, $1.20. The full bill was (8/5)($1.20) = $1.92. A) $1.80 B) $1.92 C) $2.00 D) $3.20	40. B

The end of the contest ✍ **7**

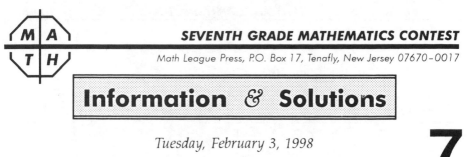
Information & Solutions

Tuesday, February 3, 1998

Contest Information

7

- **Solutions** Turn the page for detailed contest solutions (written in the question boxes) and letter answers (written in the *Answers* column to the right of each question).

- **Scores** Please remember that *this is a contest, not a test*—and there is no "passing" or "failing" score. Few students score as high as 30 points (75% correct). Students with half that, 15 points, *deserve commendation!*

- **Answers & Rating Scale** Turn to page 139 for the letter answers to each question and the rating scale for this contest.

Copyright © 1998 by Mathematics Leagues Inc.

1. $0.11 \div 1 = 0.22 \div 2 = \ldots = 0.66 \div 6 = 0.77 \div 7$.
   A) $1 \div 11$ B) $0.66 \div 6$ C) $7 \div 0.77$ D) $77 \div 7$

   1. B

2. Every student has 1 more pencil than pens. There are 1998 more pencils than pens, so there are 1998 students here.
   A) 1998   B) 3996   C) 5994   D) 9990

   2. A

3. $7/8 = 0.875$, so choice D is correct.
   A) $\frac{112}{8}$ B) $12\frac{7}{8}$ C) $13\frac{1}{8}$ D) $13\frac{7}{8}$

   3. D

4. $\frac{1}{5} + \frac{2}{10} + \frac{3}{15} + \frac{4}{20} + \frac{5}{25} = \frac{1}{5} + \frac{1}{5} + \frac{1}{5} + \frac{1}{5} + \frac{1}{5} = 1$.
   A) $\frac{1}{5}$ B) $\frac{15}{25}$ C) $\frac{35}{50}$ D) 1

   4. D

5. 1998 cm $= (1998 \div 100)$ m $= 19.98$ m.
   A) 0.1998   B) 1.998   C) 19.98   D) 199.8

   5. C

6. If $5 \times \blacklozenge = 5+4+3+2+1 = 15 = 5 \times 3$, then $\blacklozenge = 3$.
   A) 3   B) 5   C) 10   D) 15

   6. A

7. $15^4 = (3 \times 5)^4 = 3^4 \times 5^4$.
   A) $3^2 \times 5^2$ B) 3×5^4 C) $3^3 \times 5^4$ D) $3^4 \times 5^4$

   7. D

8. 3 hrs 30 mins after 8:45 A.M. is 30 mins after 11:45 A.M., which is 12:15 P.M.
   A) 11:15 A.M.   B) 11:15 P.M.
   C) 12:15 A.M.   D) 12:15 P.M.

   8. D

9. Sum $= 1 + 40 + 900 + 16\,000 = 16\,941$.
   A) 2541   B) 4321   C) 16\,940   D) 16\,941

   9. D

10. 1 million $= 10^6 = (2 \times 5)^6 = 2^6 \times 5^6$. The only primes are 2 and 5.
    A) one   B) two   C) three   D) ten

    10. B

11. Only if the dividend has an even # of 1's will 11 be a factor.
    A) 111   B) 1111   C) 111\,111   D) 11\,111\,111

    11. A

12. Since thousandths' digit is 5, round up to 19.98.
    A) 19.97   B) 19.976   C) 19.98   D) 20

    12. C

13. My town had 180 sunny days and $140 + 45 = 185$ non-sunny days. The ratio of sunny to non-sunny days is 180:185.
    A) 140:180   B) 180:140   C) 180:185   D) 180:365

    13. C

14. Dividing, $1998 \div 36$ is *not* a whole number.
    A) 37   B) 36   C) 27   D) 18

    14. B

15. Since $0.0001 = 1/10\,000$, the reciprocal is $10\,000/1$.
    A) 0.9999   B) 1.0001   C) 10\,000   D) 100\,000

    15. C

16. No odd number can have 2 as a factor.
    A) 2   B) 3   C) 5   D) 7

    16. A

Go on to the next page ⟫ **7**

17. $1/1 = 1$; $1/2 = 0.5$; $1/3 = 0.33 \ldots$; $1/4 = 0.25$; $1/5 = 0.2$, etc. A) less than or equal to 1 B) less than 1 C) greater than 1 D) greater than or equal to 1	17. A
18. Ted's and Ann's ages differ by 8 and add to 30, so Ann is 11, Ted is 19, and Bob is 14. A) 25 B) 20 C) 16 D) 14	18. D
19. $\sqrt{\frac{1}{16} + \frac{1}{16} + \frac{1}{16} + \frac{1}{16}} = \sqrt{\frac{4}{16}} = \sqrt{\frac{1}{4}} = \frac{1}{2} = 0.5$. A) 1 B) 0.5 C) 0.25 D) 0.125	19. B
20. Start at \$3.02. Subtract 29¢ until result is divisible by 31¢: 302, 273, 244, 215, <u>186</u>. OR, start with 10 plain. Change 6 for propellers to get $6 \times 2¢ = 12¢$ more. A) 3 B) 4 C) 5 D) 6	20. D
21. $\frac{2}{3}\%$ of $600 = \frac{2}{3} \times \frac{1}{100} \times 600 = \frac{2}{3} \times 6 = 4$. A) 4 B) 6 C) 40 D) 400	21. A
22. Clearly, the different one is C or D. A and B are both 1/8. A) $\frac{4^3}{2^9}$ B) $\frac{2^6}{8^3}$ C) $\frac{1}{8}$ D) $\frac{1}{4}$	22. D
23. A race is 22.5 km long. If I want to finish the race in 7.5 hours or less, my average speed *must* be $22.5/7.5 = 3$ km/hr or faster. A) 2.5 km/hr B) 3 km/hr C) 3.5 km/hr D) 4 km/hr	23. B
24. Ann is 1st, Bob 2nd, and Ray, 3rd behind Ann, is 4th. Jan and Pat together must be 5th and 6th, so Carol is 3rd, between Bob and Ray. A) Bob & Ray B) Ann & Bob C) Ray & Pat D) Ray & Jan	24. A
25. My mailman delivers mail to 7 streets in all. Since there are 210 houses on these streets altogether, the average number of houses per street is $210/7 = 30$. A) 30 B) 32 C) 34 D) 35	25. A
26. $1997^{1998} : 1997^{1997} = 1997^{1998-1997} = 1997^1$. A) 1 B) 1997 C) 1998 D) 1997^2	26. B
27. 1 is not prime. The primes could be 3, 5, and 7, whose avg is 5. A) 3 B) 5 C) 7 D) 11	27. B
28. I have 5 coins of equal value in my left pocket, and 2 coins of equal value in my right pocket. If the total value of the coins in each pocket is the same, I have 5 dimes and 2 quarters. A) pennies B) nickels C) dimes D) quarters	28. C
29. Each 3×5 photo covers area 15. Four of these would cover area 60. Since the area of the 8×12 page is 96, the area of that part of the page not covered by photos is $96 - 60 = 36$. A) 27 B) 35 C) 36 D) 81	29. C

30.	9^2 has 2 digits; 99^2 has 4 digits; $999\,999\,999\,999^2$ has $2 \times 12 = 24$ digits. A) 22 B) 23 C) 24 D) 144	30. C
31.	A ball rolled around the rim twice before falling through. The rim diameter is 50 cm, so the ball rolled 50π cm each of the two times it circled the rim, for a total of 100π cm. A) 50π cm B) 100π cm C) 150π cm D) 200π cm	31. B
32.	The sums of the 3 pairs are 1997, 1998, 1999. Add all three of these sums to get the sum of six numbers (each boy's number added twice). Now halve this to get the sum of the three numbers. A) 1998 B) 2991 C) 2997 D) 5994	32. C
33.	All sides of a 6-8-10 right triangle are integers. A) 10 B) 12 C) 16 D) 18	33. A
34.	Jane's hair grows at the rate of 0.5 cm every 4 weeks, which is $0.5/4 = 0.125$ cm per week. She cuts off 0.25 cm after the 6th week. After 11 weeks, her hair will be $50 + 11 \times 0.125 - 0.25 = 51.125$ cm long. A) 51 B) 51.125 C) 51.375 D) 51.5	34. B
35.	If the sum of the squares of two sides of a triangle equals the square of the third side, then the triangle is a right triangle. A) 45° B) 60° C) 90° D) 180°	35. C
36.	$1.998 \times 10^{20} - 1.997 \times 10^{20} = 0.001 \times 10^{20} = 10^{17}$. A) 10^{16} B) 10^{17} C) 10^{23} D) 10^{24}	36. B
37.	4 tics = 12 tacs = 15 toes. Now divide through by 15. A) $\frac{1}{15}$ B) $\frac{1}{5}$ C) $\frac{4}{15}$ D) $\frac{5}{12}$	37. C
38.	A ferris wheel ride costs 3 tickets. A carousel ride costs 2 tickets. I go on each ride at least once. If I ride the ferris wheel only twice ($6), I can ride the carousel 12 times ($24). A) 15 B) 14 C) 13 D) 12	38. D
39.	If $19/1998 <$ reciprocal $< 98/1998$, then $1998/19 \approx 20.3 <$ integer $< 1998/19 \approx 105.1$. Since the integer can be 21, 22, . . . , or 105, there are $105 - 21 + 1 = 85$ different integers. A) 78 B) 79 C) 84 D) 85	39. D
40.	Add consecutive odds. You get square of # of #s. $1+3 = 4 = 2^2$; $1+3+5 = 9 = 3^2$, etc. You need at least $\sqrt{12321} = 111$ numbers. A) 111 B) 121 C) 131 D) 221	40. A

The end of the contest 🖎 **7**

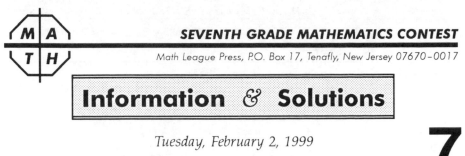

Information & Solutions

Tuesday, February 2, 1999

Contest Information

7

- **Solutions** Turn the page for detailed contest solutions (written in the question boxes) and letter answers (written in the *Answers* column to the right of each question).

- **Scores** Please remember that *this is a contest, not a test*—and there is no "passing" or "failing" score. Few students score as high as 30 points (75% correct). Students with half that, 15 points, *deserve commendation!*

- **Answers & Rating Scale** Turn to page 140 for the letter answers to each question and the rating scale for this contest.

1. Multiply by 100. Which of 1900, 1990, 2000, 2001 is nearest 1999? A) 19 B) 19.9 C) 20 D) 20.01	1. C
2. Rearranging, $2001+1998-1000-1000 = 1999 = 1999 - 0$. A) 0 B) 1 C) 1000 D) 2000	2. A
3. $\pi = 3.14159265359\ldots$, so its millionths' digit is the one 6 places to the right of the decimal point; it's a 2. A) 2 B) 5 C) 6 D) 9	3. A
4. $\sqrt{1\times9\times9\times9} = 1\times3\times3\times3 = 27$. A) 9 B) 19 C) 27 D) 99	4. C
5. $(1024\div64)\div8\div2 = (16)\div8\div2 = 2\div2 = 1$. A) 1 B) 2 C) 4 D) 16	5. A
6. $(2\text{ m}) - (1000\text{ mm}) = 2\text{ m} - 1\text{ m} = 1\text{ m} = 100\text{ cm}$. A) 10 cm B) 100 cm C) 20 cm D) 200 cm	6. B
7. 1% of 1 = 1/100, which is *not* equal to 1. A) 1% of 100 B) $100\times1\%$ C) $1\times100\%$ D) 1% of 1	7. D
8. Twelve hrs and one min before midnight is 1 min before noon. A) 12:01 P.M. B) 12:01 A.M. C) 11:59 P.M. D) 11:59 A.M.	8. D
9. $6\times(18) = 6\times(2\times9) = (6\times2)\times9 = 12\times9$. A) 5 B) 9 C) 11 D) 12	9. B
10. $10\,499/1000 = 10.499$, which is nearer to 10 than 11. A) 1 B) 1.1 C) 10 D) 11	10. C
11. 100% of 50 is $50 = 200\%$ of 25. A) 25 B) 75 C) 100 D) 150	11. A
12. $\sqrt{9}+\sqrt{4}+\sqrt{9}+\sqrt{4} = 3+2+3+2 = 10 = \sqrt{100}$. A) $\sqrt{10}$ B) $\sqrt{26}$ C) $\sqrt{36}$ D) $\sqrt{100}$	12. D
13. At the Founder's Day Gift Wrap Sale, I can get 40 gifts wrapped for \$2, since 200 cents $\div$ 5¢ per gift = 40 gifts. A) 10 B) 20 C) 40 D) 100	13. C
14. $1\frac{1}{2}\times1\frac{1}{3}\times1\frac{1}{4}\times1\frac{1}{5} = \frac{3}{2}\times\frac{4}{3}\times\frac{5}{4}\times\frac{6}{5} = \frac{6}{2} = 3$. A) 2 B) 3 C) 6 D) 12	14. B
15. If its radius were 7.5, then its diameter would be 15, an integer. A) 7.25 B) 7.5 C) 7.75 D) π	15. B

Go on to the next page ⇒ **7**

16. $2^2 \times 3^3 \times 4^4 = 4^1 \times 3^3 \times 4^4 = 3^3 \times 4^5$.

    A) 4   B) 4^3 C) 4^4 D) 4^5

    16. D

17. The largest prime factor of
    $26\,000\,000\,000 = 2 \times 13 \times 10^9 = 13 \times 2^{10} \times 5^9$ is 13.

    A) 2   B) 5   C) 13   D) 26

    17. C

18. 200% of $1 = 2 \times 1 = 2$, a prime.

    A) 1   B) 2   C) 3   D) 5

    18. A

19. 5% = 0.05 is half (50%) of 0.10 = 0.1.

    A) 10%   B) 50%   C) 200%   D) 500%

    19. B

20. The number of 2 cm × 2 cm tiles (area 4 cm^2) needed to cover a 40 cm × 50 cm region (area 2000 cm^2) is 2000 ÷ 4 = 500.

    A) 500   B) 1000   C) 2000   D) 8000

    20. A

21. 1 penny + 1 nickel + 1 dime = 16¢. It takes 15 of these 16¢ groups to be worth $2.40. The value of the 15 nickels is 75¢.

    A) 45 cents   B) 50 cents   C) 60 cents   D) 75 cents

    21. D

22. My hair must grow 5 cm. Since 5 cm ÷ 0.5 cm = 10, the answer is C.

    A) 25   B) 16   C) 10   D) 6

    22. C

23. My trophy holds 1ℓ. I drink 2/3 ℓ, leaving 1/3 ℓ. Replacing $3/4 \times 2/3\ \ell = 1/2\ \ell$ leaves 1/3 + 1/2 = 5/6 ℓ of water in my trophy.

    A) $\frac{1}{2}\ell$ B) $\frac{2}{3}\ell$ C) $\frac{3}{4}\ell$ D) $\frac{5}{6}\ell$

    23. D

24. There are seven 7^2 terms, which is 7×7^2.

    A) 7^2 B) 7^3 C) 7^7 D) 7^8

    24. A

25. $1 - \frac{6}{18} - \frac{3}{18} - \frac{2}{18} - \frac{1}{18} = 1 - \frac{12}{18} = \frac{6}{18} = \frac{1}{3}$.

    A) $\frac{1}{6}$ B) $\frac{2}{9}$ C) $\frac{1}{3}$ D) $\frac{1}{2}$

    25. C

26. If 2 quadrilaterals share a side, the result has at least 4 sides.

    A) a hexagon   B) a pentagon   C) a square   D) a triangle

    26. D

27. If the lines are parallel, $x° + y° = 45° + 45° = 90°$.

    A) 80°   B) 90°   C) 100°   D) 110°

    27. B

28. C ≈ 1/10 and D ≈ 1/10, so B is smallest.

    A) $\frac{1}{9}$ B) $\frac{1}{19}$ C) $\frac{19}{199}$ D) $\frac{199}{1999}$

    28. B

Go on to the next page ⫸ **7**

29. A square of side 3 is cut from a square of side 9 as shown. Area of shaded region = $9^2 - 3^2 = 81 - 9 = 72$.

   A) 6   B) 24   C) 64   D) 72

   29. D

30. The lcm of (2×3), (3×5), and (2×5) is $2 \times 3 \times 5 = 30$.

   A) 1   B) 20   C) 30   D) 900

   30. C

31. Three $\cong$ $\bigodot$s are tangent, so each side of the triangle is 10 and each radius is 5. The area of each circle is 25π.

   A) 25π B) 36π C) 75π D) 100π

   31. A

32. From noon one day until noon the next, it passes twice: at 1 P.M. and at 1 A.M. In 2 days, it passes 4 times.

   A) 2   B) 3   C) 4   D) 48

   32. C

33. $0.9999 + \underline{?} = 1$, so $\underline{?} = 1 - 0.9999 = 0.0001$.

   A) 0.00001   B) 0.0001   C) 0.0111   D) 0.1111

   33. B

34. Two consecutive odd numbers never have any positive common divisor except 1. Thus, the number of days that Weird Uncle Barney could stay was 1.

   A) 1   B) 2   C) 3   D) 7

   34. A

35. Ones' digits: 2,4,8,6,2,4,8,6,...; 2^{1999} ends in 8.

   A) 2   B) 4   C) 6   D) 8

   35. D

36. If the $\angle$'s of a $\triangle$ have whole-number degree measures, and one $\angle$ is 80°, smallest could be 1°. The third $\angle$ is $180° - 80° - 1° = 99°$.

   A) 79°   B) 80°   C) 99°   D) 100°

   36. C

37. If all are parallel, there are 0 crossings. If all pass through one point, there is 1 crossing. If 9 are parallel and 1 is not, there are 9 crossings.

   A) 0   B) 1   C) 2   D) 9

   37. C

38. An equilateral $\triangle$ shares a side with an isosceles right $\triangle$ as shown. $m\angle ABC = 45° + 60° = 105°$.

   A) 105°   B) 120°   C) 135°   D) 150°

   38. A

39. 1997 & 1999 are prime; 1998's prime factors are 2, 3, 37; 2000's prime factors are 2 and 5. A) 1997 B) 1998 C) 1999 D) 2000

   39. B

40. **1,1,1,2,1,3,1,4,1,**

   2nd #: 1; 4th #: 2; 6th #: 3; 8th #: 4; . . . ; 200th #: 100.

   A) 1   B) 100   C) 101   D) 200

   40. B

The end of the contest ✍ **7**

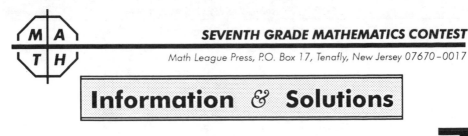

Information & Solutions

February, 2000

Contest Information

7

- **Solutions** Turn the page for detailed contest solutions (written in the question boxes) and letter answers (written in the *Answers* column to the right of each question).

- **Scores** Please remember that *this is a contest, not a test*—and there is no "passing" or "failing" score. Few students score as high as 30 points (75% correct). Students with half that, 15 points, *deserve commendation!*

- **Answers & Rating Scale** Turn to page 141 for the letter answers to each question and the rating scale for this contest.

1. $111\,111 + 333\,333 + 222\,222 + 444\,444 = 111\,111 \times (1+2+3+4) = 222\,222 \times 5$.
   A) 1   B) 4   C) 5   D) 10   **1. C**

2. $257 \div 5 = 51.4$, and $.4 = 4/10 = 2/5$, signifying a remainder of 2.
   A) $257 \div 5$ B) $228 \div 6$ C) $195 \div 3$ D) $176 \div 4$ **2. A**

3. Use commutativity: 6 twos = 2 sixes, and 8 threes = 3 eights.
   A) 3   B) 6   C) 8   D) 12   **3. A**

4. $2 + (10 \times 2) + (100 \times 2) + (1000 \times 2) = 2 + 20 + 200 + 2000 = 2222$.
   A) 224   B) 2000   C) 2220   D) 2222   **4. D**

5. Only choice A is a whole number, with $182 \div 7 = 26$.
   A) $\frac{182}{7}$ B) $\frac{172}{12}$ C) $\frac{189}{17}$ D) $\frac{178}{21}$ **5. A**

6. Four 2^2s $= 2^2 \times 4 = 2^2 \times 2^2$.
   A) 2^1 B) 2^2 C) 2^3 D) 2^4 **6. B**

7. The tens' digit is second to the left of the decimal point. The tens' digit of choice A is a 3, the largest tens' digit below.
   A) \$1231.21   B) \$1123.03   C) \$3010.30   D) \$2302.12   **7. A**

8. February 1st, 8th, 15th, 22nd, and 29th can all fall on a Tuesday.
   A) 3   B) 4   C) 5   D) 6   **8. C**

9. $\frac{1+2}{3} + \frac{4+5}{6} = \frac{3}{3} + \frac{9}{6} = \frac{6}{6} + \frac{9}{6} = \frac{15}{6} = \frac{7+8}{9-3}$.
   A) 0   B) 3   C) 6   D) 12   **9. B**

10. A square with side 9 has area 81. A rectangle with width 3 and area 81 has length $81 \div 3 = 27$. Its perimeter is $3+27+3+27 = 60$.
    A) 27   B) 36   C) 60   D) 81   **10. C**

11. The product of 2 positive numbers is always positive.
    A) greater than 0   B) greater than 1
    C) greater than 2   D) at least 2   **11. A**

12. $98 \div 24 = 4\,2/24 = 4$ days + 2 hr. after 11 P.M.
    Sun. = 2 hr. after 11 P.M. Thurs. = 1 A.M. Fri.
    A) Tues.   B) Wed.   C) Thurs.   D) Fri.   **12. D**

13. $4+9+25 = 38 = 2 \times 19$, and 19 is a prime.
    A) 5   B) 13   C) 19   D) 37   **13. C**

14. The smallest prime is 2. Its reciprocal is $\frac{1}{2}$.
    A) 0   B) $\frac{1}{2}$ C) 1   D) 2   **14. B**

15. $(20 \times 100) - (20 \times 10) - (20 \times 1) = 20 \times (100 - 10 - 1)$.
    A) 20×111 B) 20×109 C) 20×91 D) 20×89 **15. D**

Go on to the next page ⇒ **7**

16. Days: $4.5 \div 24 = 3/16$; mins.: $4.5 \times 60 = 270$; secs.: $270 \times 60 = 16200$. A) $\frac{3}{16}$ day B) 270 min C) 16 200 sec D) $\frac{3}{100}$ week	16. D
17. Since 1 m = 100 cm, 15 cm is $15/100 = 3/20$ of 1 m. A) $\frac{1}{10}$ B) $\frac{3}{20}$ C) $\frac{1}{15}$ D) $\frac{10}{15}$	17. B
18. One thousandth = 0.001. Multiply by 100 to get = 0.100 = 0.1. A) 100 B) 1000 C) 10 000 D) $\frac{1}{100}$	18. A
19. Since 40% + 38% = 78% answered "yes" or "no," 100% − 78% = 22% said "Ouch." Finally, 22% of 250 = $0.22 \times 250 = 55$. A) 22 B) 25 C) 55 D) 195	19. C
20. Subtract first, so $\sqrt{25 - 16} = \sqrt{9} = 3$. A) 9 B) 3 C) 1 D) −11	20. B
21. Jill is 3 years older than Jack, so Jack is 20 and Jill is 23. In 2 years, Jill will be 25. A) 20 B) 22 C) 23 D) 25	21. D
22. My average on 5 tests was 95. Their sum was $5 \times 95 = 475$. A) 95 B) 100 C) 475 D) 495	22. C
23. Add first: $1 + \frac{7}{8} = \frac{15}{8}$. The reciprocal of this is $\frac{8}{15}$. A) $\frac{8}{15}$ B) $1 + \frac{8}{7}$ C) $\frac{15}{8}$ D) $\frac{7}{15}$	23. A
24. If I have 1 dime and 2 quarters, I have 10¢ in dimes and 50¢ in quarters. Notice that I have one-half as many dimes as quarters. A) one-half B) one-fifth C) two-thirds D) twice	24. A
25. The year 2000 is a leap year, so the average is $366 \div 12 = 30.5$. A) 29 B) 30 C) 30.5 D) 31	25. C
26. 162 is divisible by 1, 2, 3, 6, and 9, but *not* by 0, 4, 5, 7, or 8. A) 6 B) 5 C) 4 D) 3	26. B
27. Filling both blanks with the same choice, try to satisfy "_?_ months + 55 = _?_ years." With choice C, we get: 60 mos. + 55 yrs = 60 yrs. A) 48 B) 55 C) 60 D) 66	27. C
28. 30 = 10% of 300 = 1% of 3000 = 0.1% of 30 000. A) 3 B) 300 C) 3000 D) 30 000	28. D
29. midway # = $(1 234 567 + 7 654 321) \div 2 = 8 888 888 \div 2 = 4 444 444$. A) 3 765 432 B) 4 321 321 C) 4 444 444 D) 3 456 789	29. C

Go on to the next page ⟩⟩⟩➡ **7**

30. If $P = 3$ cm, then $s = 3/4$ cm, and area $= (3/4)^2 = 9/16$ cm^2. A) $\frac{9}{16}$ cm^2 B) $\frac{3}{2}$ cm^2 C) 3 cm^2 D) 9 cm^2	30. A
31. Squares of consecutive integers differ by the sum of the integers. A) 0 B) 34592867542 C) 34592867543 D) 34592867544	31. D
32. 10 oranges cost as much as 4 grapefruits, and 4 grapefruits cost as much as 12 apples, so 10 oranges cost as much as 12 apples. A) 6 B) 9 C) 10 D) 12	32. D
33. If the area of the "other" circle is $\pi(r)^2$, then the area of the *Winner's Circle* is $4\pi r^2 = \pi(2r)^2$; so its radius is twice the "other" circle's radius. A) exactly B) twice C) one-fourth D) 4 times	33. B
34. A ≈ 0.7071; B $= 0.5$; C $=$ D $= 0.25$. A) $\sqrt{\frac{1}{2}}$ B) $\sqrt{\frac{1}{4}}$ C) $\frac{1}{4}$ D) $\left(\frac{1}{2}\right)^2$	34. A
35. The second hand makes 1 revolution per minute $= 60 \times 24$ per day. A) 60 B) 1440 C) 3600 D) 86 400	35. B
36. The digits of 85 030 add up to 16, so 3 is not a factor of 85 030. A) 543 420 B) 85 030 C) 72 630 D) 53 430	36. B
37. 30 days: 4 mos.; 31 days: 7 mos.; 29 days: 1 mo.; so there are 187 odd days and 179 even. I save \$187 + \$2 × 179 = \$545. A) \$16 B) \$538 C) \$545 D) \$549	37. C
38. $\sqrt{999\,999\,999} \approx 31622.7$, so largest square is $31622^2 = 999\,950\,884$. A) 0 B) 1 C) 4 D) 9	38. C
39. The five 1 km legs of the race took $(1\div5 + 1\div4 + 1\div3 + 1\div2 + 1\div1)$ hours $= (12 + 15 + 20 + 30 + 60)$ minutes $= 137$ minutes. A) 120 B) 137 C) 216 D) 685	39. B
40. Each piece has straight cuts as edges, so each is a polygon and each has *at least* 3 edges. Altogether, the three polygons have a total of *at least* $3 \times 3 = 9$ edges. See the illustrations below. A) 8 B) 9 C) 10 D) 11	40. A

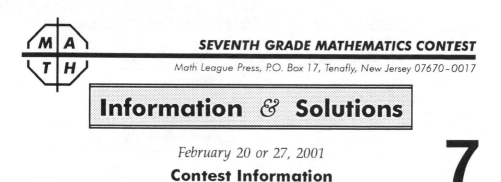

Information & Solutions

February 20 or 27, 2001

Contest Information

7

- **Solutions** Turn the page for detailed contest solutions (written in the question boxes) and letter answers (written in the *Answers* column to the right of each question).

- **Scores** Please remember that *this is a contest, not a test*—and there is no "passing" or "failing" score. Few students score as high as 30 points (75% correct). Students with half that, 15 points, *deserve commendation!*

- **Answers & Rating Scale** Turn to page 142 for the letter answers to each question and the rating scale for this contest.

		Answers
1. $44\,444 \times 2 = 88\,888 = 80\,000 + 8888$. A) 0 　　　B) 8000 　　　C) 8080 　　　D) 8888		1. D
2. The hundredths' digit is a 9, so round the tenths' digit up. A) 89.9 　　　B) 89.99 　　　C) 90 　　　D) 90.0		2. A
3. First multiply, then add: $1000 + 0 \times 0 + 1000 = 1000 + 0 + 1000$. A) 0 　　　B) 1000 　　　C) 2000 　　　D) 1 000 000		3. C
4. The weight of 3 tubas is 3 times the weight of 3 trumpets, so each tuba weighs 3 times as much as each trumpet. Thus, multiply by 3. A) 1 　　B) 2 　　C) 3 　　D) one-third		4. C
5. $8888 = 9999 - 1111$. A) 177 　B) 999 　C) 1000 　D) 1111		5. D
6. Add digits. 3 divides A & D. GCF with A is 9. A) 999 　　B) 1999 　　C) 2999 　　D) 3999		6. D
7. To get as many factors as possible, make every factor a 2. How many? Try 10 twos: $2 \times 2 \times 2 \times 2 \times 2 \times 2 \times 2 \times 2 \times 2 \times 2 = 1024$, while with 11 twos, we get $2 \times 2 \times 2 \times 2 \times 2 \times 2 \times 2 \times 2 \times 2 \times 2 \times 2 = 2048$. A) 6 　　　B) 7 　　　C) 11 　　　D) 12		7. C
8. Since twice 12 equals 24, we take half of 12 to get 6. A) 3 　　　B) 6 　　　C) 12 　　　D) 48		8. B
9. $1 \div \frac{1}{10} = 1 \times 10 = 10 \div 1$. A) 100 　B) 10 　C) 1 　D) $\frac{1}{100}$		9. C
10. Perimeter = 24, so length + width = 12. Length is 2 more than width, so length = 7, width = 5, area = 35. A) 27 　B) 35 　C) 42 　D) 48		10. B
11. $3^2 + 6^2 = 45 = 9^2 - 6^2$. A) 6^2 　　　B) 3^2 　　　C) 2^2 　　　D) 0^2		11. A
12. Working *inside* parentheses first: A) 2×4 B) 4^2 C) 4^2 D) 4×4. A) $2 \times (2^2)$ 　　B) $(2 \times 2)^2$ 　　C) $(2 + 2)^2$ 　　D) $2^2 \times 2^2$		12. A
13. Drop zeros: $4 \times 5 \times 6 \times 7 = 840$. Add $8+4+0 = 12$. It's divisible by 4. A) 4 　　　B) 5 　　　C) 7 　　　D) 11		13. A
14. $250\% = 2.5$, while choice D $= (0.5)^2 = 0.5 \times 0.5 = 0.25 < 250\%$. A) $\sqrt{6.25} = 2.5$ B) $\frac{5}{2} = 2.5$ 　C) 2.5 　　　D) $(0.5)^2$		14. D
15. 1 day = 24 hrs = 60×24 mins = 1440 mins = 60×1440 seconds. A) $\frac{1}{12}$ 　　B) $\frac{1}{24}$ 　　C) $\frac{1}{60}$ 　　D) $\frac{1}{3600}$		15. C

Go on to the next page ▮▮▮➡ **7**

16. Multiply first, then add: A) 6+1 = 7 B) 3+2 = 5 C) 6 D) 6. A) $3 \times 2 + 1$ B) $3 + 1 \times 2$ C) $1 \times 3 \times 2$ D) $2 + 3 + 1$	16. A
17. 1) $-6°$ to $0°$ is a $6°$ rise; 2) $0°$ to $19°$ is a $19°$ rise; 3) $6° + 19° = 25°$. A) -13 B) 13 C) 25 D) 27	17. C
18. $3 \text{ cm} \times (3 \text{ cm}) = 3 \text{ cm} \times (3 \times 1 \text{ cm}) = (3 \text{ cm} \times 3) \times 1\text{cm} = (3 \times 3 \text{ cm}) \times 1$ cm. A) 1 cm B) 1 cm^2 C) 1 D) 0 cm	18. A
19. $(1 \times 10) + (5 \times 1) + (15 \times 0.1) = 10 + 5 + 1.5 = 16.5$. A) 15.15 B) 15.5 C) 16.15 D) 16.5	19. D
20. They popped 31 balloons. Pat popped fewer. Use "guess and check." Try 16 & 15, then 17 & 14, then 18 and 13. Pete popped 18. Pat popped 13. A) 5 B) 13 C) 18 D) 26	20. B
21. The reciprocal of $\frac{4}{3}$ is $\frac{3}{4} = \frac{9}{12} = \frac{12}{16}$. A) $\frac{8}{12}$ B) $\frac{12}{16}$ C) $\frac{16}{12}$ D) $\frac{24}{16}$	21. B
22. 100 quarters $= 100 \times 25¢ = 2500¢$. The others are worth 2000¢ each. A) 200 dimes B) 400 nickels C) 2000 pennies D) 100 quarters	22. D
23. $\sqrt{49} + \sqrt{81} = 7 + 9 = 4^2$. A) 2^2 B) 4^2 C) 8^2 D) 16^2	23. B
24. The sum of 8 numbers whose average is 13 is $8 \times 13 = 104$. A) 13 B) 21 C) 64 D) 104	24. D
25. In dollars, choice B $= 0.35 \times 0.70¢ = 0.245¢$, less than a quarter. A) $2 \times 35¢ \times 35$ B) $0.35 \times 70¢$ C) $7 \times \$3.50$ D) $\$7 \times 3.5$	25. B
26. 3.5 is 0.5 more than the odd prime number 3. Note: 1 is not prime. A) 3.5 B) 2.5 C) 1.5 D) 0.5	26. A

27. Whenever February has 5 Sundays, Feb. 1 falls on a Sunday and Feb. 14 falls on a Saturday.
  A) Thursday   B) Friday
  C) Saturday   D) Sunday
	27.
	C

28. The sum of the square roots of 0, 0, 0, 0, and 25 is 5. Choose other numbers? The sum is larger! A) less than 5 B) equal to 5 C) at least 5 D) at most 5	28. C
29. $\left(\frac{2}{5} \div \frac{5}{2}\right) = \frac{4}{25} = 16\%$. A) 4 B) 16 C) 20 D) 40	29. B
30. Choice A is divisible by 3; choices C and D are even and > 2. A) $12^2 + 33^2$ B) $13^2 + 38^2$ C) $14^2 + 18^2$ D) $15^2 + 49^2$	30. B

Go on to the next page ‖▶ **7**

31. Calendars advance 1 day in a reg. year and 2 in a leap yr. Eye could not be 6. That would require 7 consecutive years without a leap yr.

    A) 6   B) 7   C) 8   D) 13

    31. A

32. Try $1+2+3+4+5+6+7+8+9+10 = 55$. When 55 is divided by 10, remainder is 5.

    A) 0   B) 1   C) 5   D) 9

    32. C

33. $\sqrt{(1 \times 2 \times 3 \times 4 \times 5 \times 6) \times 7 \times (8 \times 9 \times 10)} = \sqrt{720 \times 720 \times 7}$.

    A) $30 + \sqrt{7}$ B) $30 \times \sqrt{7}$ C) $720 + \sqrt{7}$ D) $720 \times \sqrt{7}$

    33. D

34. Longest side < sum of 2 smaller sides, so longest side < 9. It can be 8, as illustrated by (8,8,2), (8,7,3), (8,6,4), and (8,5,5).

    A) 7   B) 8   C) 9   D) 10

    34. B

35. Lee runs at the rate of $\frac{3}{4}$ km/min, so Lee runs 3 km in 4 mins. Pat runs 5 km in (4+4) mins = 8 mins, so Pat's rate is 5 km/8 mins.

    A) $\frac{5}{8}$ km/min   B) $\frac{4}{3}$ km/min   C) $\frac{5}{4}$ km/min   D) 2 km/min

    35. A

36. $2^{2000} + 2^{2000} =$ twice $2^{2000} = 2^1 \times 2^{2000} = 2^{2001}$.

    A) 2   B) 4   C) 2000   D) 2^{2000}

    36. D

37. $? = (1 + \frac{2}{3+4}) \times (2 + \frac{3}{4+5}) = \frac{9}{7} \times \frac{7}{3}$.   A) 1   B) 2   C) 3   D) 4

    37. C

38. The first fold could make the dimensions 2 by 4. The second fold could then make the dimensions 1 by 4 or 2 by 2.

    A) 4   B) 3   C) 2   D) 1

    38. B

39. The min. hand moves 6° each min. The hour hand moves 1/12 as fast, 1/2° per min. At 12:12, the angle between the hands is $72° - 6° = 66°$.

    A) 48°   B) 60°   C) 66°   D) 72°

    39. C

40. Sum is 5000, so 1st digit isn't 5 or 8. Last digit can't be 2 or 8: the mirror image would then exceed 5000. The possible mirror image number pairs are (1285,2851), (1825,2581), and (2185,2815). The pair whose sum is 5000 is 2185 and 2815. Their difference is $2815 - 2185 = 630$.

    A) 566   B) 630   C) 1566   D) 1866

    40. B

The end of the contest   7

8th Grade Solutions

1996-1997 through 2000-2001

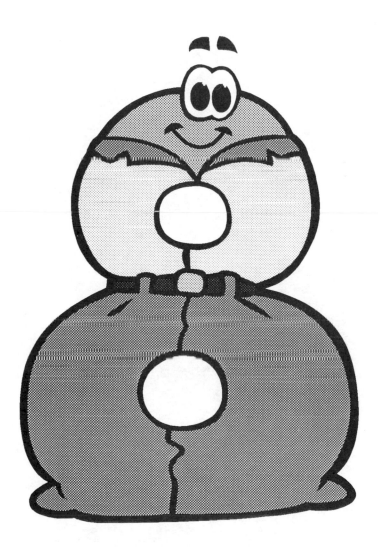

EIGHTH GRADE MATHEMATICS CONTEST

Math League Press, P.O. Box 17, Tenafly, New Jersey 07670-0017

Information & Solutions

Tuesday, February 4, 1997

Contest Information

8

- **Solutions** Turn the page for detailed contest solutions (written in the question boxes) and letter answers (written in the *Answers* column to the right of each question).

- **Scores** Please remember that *this is a contest, not a test*—and there is no "passing" or "failing" score. Few students score as high as 30 points (75% correct). Students with half that, 15 points, *deserve commendation!*

- **Answers & Rating Scale** Turn to page 143 for the letter answers to each question and the rating scale for this contest.

1. Subtracting a negative # is the same as adding a positive #. A) even B) odd C) positive D) zero	1. C
2. $3 \div \left(\frac{1}{7} + \frac{2}{7} + \frac{3}{7} + \frac{4}{7} + \frac{5}{7} + \frac{6}{7} \right) = 3 \div 3 = 1.$ A) 1 B) $\frac{3}{7}$ C) 3 D) $\frac{7}{3}$	2. A
3. Fifty of Pat's paper airplanes landed in the wastebasket. If Lee was only 80% as effective with wads of paper, only $0.8 \times 50 = 40$ of Lee's wads landed in the wastebasket. A) 32 B) 40 C) 50 D) 60	3. B
4. Every negative number is smaller than 0. A) 0 B) –0.5 C) –1 D) –10%	4. A
5. The reciprocals of 1 and –1 are 1 and –1. A) 0 B) 1 C) 2 D) 3	5. C
6. 10 AM – 24 hrs = (10 AM – 11 hrs) – 13 hrs = present – 13 hrs. A) 10 B) 11 C) 13 D) 24	6. C
7. $46 \div 6 = 7.666 \ldots$ A) $4\frac{1}{6} \approx 4.16$ B) $7\frac{2}{3} \approx 7.66$ C) $7\frac{4}{6} \approx 7.66$ D) $\frac{46}{6} \approx 7.66$	7. A
8. $0.123456789 + 0.987654321 = 1.111111110$. There are nine 1's. A) 1 B) 7 C) 8 D) 9	8. D
9. the least pos integer – the largest neg integer = $1 - (-1) = 2$. A) 2 B) 1 C) 0 D) –1	9. A
10. $(2112 - 2002) \div 11 = 110 \div 11.$ A) 2110 B) 110 C) 11 D) 10	10. B
11. If a box of 12 candles costs \$4.20, and a box of 6 candles costs \$2.40, then 42 candles (3 boxes of 12 and 1 box of 6) can cost $3 \times \$4.20 + \$2.40 = \$12.60 + \$2.40 = \$15.$ A) \$14.70 B) \$15 C) \$15.60 D) \$16.80	11. B
12. Five-hundredths = 1/20 = 50% of 1/10. A) 2 B) 5 C) 20 D) 50	12. D
13. Order of operations: $-3^2 = -(3^2) = -9$. A) 6 B) –6 C) 9 D) –9	13. D
14. By calculator, $1997 \div 1996 \approx 1.0005$. Now *round*. A) 1.01 B) 1.001 C) 1.005 D) 1.0005	14. B
15. A pentagon has $p = 5$ sides. A trapezoid has $t = 4$ sides. $p \times t = 20$. A) 15 B) 18 C) 20 D) 24	15. C

Go on to the next page ⟫ **8**

16. $2\times4\times6\times8\times10 = 2\times(1\times2\times3\times4\times5)$, so gcf $= 1\times2\times3\times4\times5 = 120.$

    A) 2   B) 4   C) 8   D) 120

16. D

17. $25\times10^8 = 2.5\times10^9 = 0.25\times10^{10} = 0.025\times10^{11}.$

    A) 25×10^8 B) 250×10^8 C) 0.25×10^{10} D) 0.025×10^{11}

17. B

18. It's now 2 years later, so reduce Jan's age advantage by 2 years. Right now, Jan is only $3 - 2 = 1$ year older than Dale.

    A) 1 year   B) 2 years   C) 3 years   D) 5 years

18. A

19. $\frac{21}{1000} + \frac{3}{100} + \frac{4}{10} = \frac{21}{1000} + \frac{30}{1000} + \frac{400}{1000} = \frac{451}{1000}.$

    A) 0.28   B) 0.2134   C) 0.4321   D) 0.451

19. D

20. My tea, with 6% tax, costs $2.65. Divide by $106\% = 1.06$ to determine cost of the tea alone.

    A) $1.65   B) $2.25   C) $2.50   D) $2.53

20. C

21. Multiply before adding; $4+16+16+4 = 40.$

    A) 40   B) 64   C) 84   D) 148

21. A

22. There are two possibilities: 50°, 50°, 80° or 50°, 65°, 65°. In each case, the smallest angle is a 50° angle.

    A) 25°   B) 40°   C) 50°   D) 80°

22. C

23. Each week, 7×30 mins (or 3.5 hrs) at 5 km/hr $= 3.5\times5$ km/week.

    A) 2.5 km   B) 17.5 km   C) 35 km   D) 150 km

23. B

24. If every side is a multiple of 3, the sum is a multiple of 3. When you divide by 3, you'll always get a whole number.

    A) an odd   B) an even   C) a whole   D) a prime

24. C

25. If a month's first and last days fall on the same day of the week, the last date is $28 + 1 = 29$, the number of days.

    A) 28   B) 29   C) 30   D) 31

25. B

26. 0.75 is the additive inverse of $-0.75 = -3/4.$

    A) $-\frac{3}{4}$ B) $\frac{3}{4}$ C) $-\frac{4}{3}$ D) $\frac{4}{3}$

26. A

27. Of every 4 kg, the couch gets 1 and the potato gets 3, a 1 to 3 ratio. Therefore, the potato weighs 3/4 of 720 kg = 540 kg.

    A) 540 kg   B) 480 kg   C) 240 kg   D) 180 kg

27. A

28. Numbers bigger than 1 have reciprocals smaller than 1.

    A) positive   B) negative   C) less than 1   D) more than 1

28. D

29. In a right triangle, the right angle is 90°, and the sum of the other two angles is also 90°.

    A) acute   B) obtuse   C) isosceles   D) right

29. D

Go on to the next page IIII➡ **8**

30. The number of spots on a dalmatian is a whole number, so the number of spots on a leopard is the square of a whole number.
    A) $121 = 11^2$ B) $144 = 12^2$ C) 164   D) $169 = 13^2$

30. C

31. The square's side is 6 and its diagonal is $\sqrt{6^2+6^2} = \sqrt{72}$. The area of the circle is $\pi r^2 = \pi(\sqrt{72}/2)(\sqrt{72}/2) = 72\pi/4 = 18\pi$.
    A) 72π B) 36π C) 24π D) 18π

31. D

32. With as many pennies as nickels, my total is 6¢ or 12¢ or 18¢ or 24¢ or any multiple of 6¢.
    A) $0.50   B) $1.00   C) $1.50   D) $1.75

32. C

33. For every $100 it costs now, it will cost $110 next year. The 2nd year's cost is 110%($110) = $121. Finally, $121/$100 is 121%.
    A) 120   B) 121   C) 130   D) 230

33. B

34. Every term after the first two terms is the sum of the two terms that precede it. This sequence begins 0.1, 0.1, 0.2 (= 0.1 + 0.1), 0.3 (= 0.1 + 0.2). The next term is 0.2 + 0.3 = 0.5.
    A) 0.3   B) 0.4   C) 0.5   D) 0.6

34. C

35. We need whole numbers that satisfy $a^2+b^2 = 26^2$. The smallest choice that works is $10^2+b^2 = 26^2$ ($b = 24$), so A is the answer.
    A) 10 cm   B) 12 cm   C) 13 cm   D) 24 cm

35. A

36. $\frac{1+1996}{1997} + \frac{2+1995}{1997} + \frac{3+1994}{1997} + \ldots$ Each of these 998 sums equals 1.
    A) 998   B) 998.5   C) 1996   D) 1997

36. A

37. $(5\times2)\times10\times(15\times6)\times20\times(25\times4)$ ends in $1+1+1+1+2 = 6$ zeros.
    A) 3   B) 4   C) 5   D) 6

37. D

38. Since $4^{10} = 2^{20}$, and 2 is a prime, the whole number divisors are the powers of 2 from 2^0 through 2^{20}. That's a total of 21 divisors. (Early Bird feasted!)
    A) 10   B) 11   C) 20   D) 21

38. D

39. $(100-99)+\ldots+(2-1)$. All 50 pairs are 1.
    A) 49   B) 50   C) 99   D) 100

39. B

40. Up to p.99, the count is 189 = 9 1-digit #s + 90 2-digit #s. New digits occur 3 at a time. The digit-count can = $189+3\times603 = 1998$.
    A) 1997   B) 1998   C) 1999   D) 2000

40. B

The end of the contest   **8**

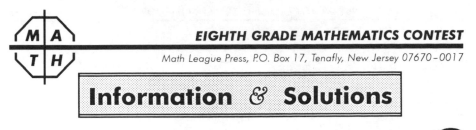
Information & Solutions

Tuesday, February 2, 1998

Contest Information

8

- **Solutions** Turn the page for detailed contest solutions (written in the question boxes) and letter answers (written in the *Answers* column to the right of each question).

- **Scores** Please remember that *this is a contest, not a test*—and there is no "passing" or "failing" score. Few students score as high as 30 points (75% correct). Students with half that, 15 points, *deserve commendation!*

- **Answers & Rating Scale** Turn to page 144 for the letter answers to each question and the rating scale for this contest.

1. Sum = $(750) + (750 + 750 + 750) = 750 \times 4$.
   A) 2   B) 3   C) 4   D) 6

   1. C

2. \$1.80 = 15 pennies + (165/5) nickels; 165/5 = 33.
   A) 25   B) 30   C) 33   D) 35

   2. C

3. Sum = $(100 + 10 + 1) \times 111 = 111 \times 111$.
   A) 10   B) 11   C) 110   D) 111

   3. D

4. Answer must be largest choice! Or, $\frac{5}{16} = 0.3125$.
   A) 0.4   B) 0.3   C) 0.25   D) 0.2

   4. A

5. By the order of operations, $96 - [(48 \div 2) \times 4] = 96 - 96 = 0$.
   A) 96   B) 90   C) 6   D) 0

   5. D

6. $1 \times 100 \times (2 \times 50) \times (4 \times 25) \times (5 \times 20) = 1 \times 100 \times 100 \times 100 \times 100 = 100^4$.
   A) 400   B) 3×100 C) 100×100 D) 100^4

   6. D

7. *Kilo* = thousand. 1 m = 1/1000 = 1/10 of 1/100 = (1/10)% of 1 km.
   A) 0.5   B) 0.1   C) 0.01   D) 0.001

   7. B

8. $(1/3 \times 3/4 \times 2)\ell = (2/4)\ell = 0.5\ell = 500$ ml.
   A) 400 ml   B) 500 ml   C) 667 ml   D) 1500 ml

   8. B

9. $1122 = 11 \times 102$, $2233 = 11 \times 203$, and $3344 = 11 \times 304$.
   A) 9   B) 10   C) 11   D) 12

   9. C

10. Huck is twice as old as Becky and 2 years older than Tom. Becky is 12, so Huck is 24 and Tom is 22.
    A) 14   B) 20   C) 22   D) 24

    10. C

11. If $\frac{3}{4} = \underline{\ ?\ } \times \frac{4}{3}$, then $\underline{\ ?\ } = \frac{3}{4} \div \frac{4}{3} = \frac{3}{4} \times \frac{3}{4} = \frac{9}{16}$.
    A) $\frac{9}{16}$ B) $\frac{3}{4}$ C) $\frac{1}{3}$ D) -1

    11. A

12. $7 \times 24 \times 60 \times 60 + 7 \times 60 \times 60 + 7 \times 60 + 7$ seconds = 630 427 seconds.
    A) 10 507   B) 605 647   C) 630 427   D) 630 440

    12. C

13. The ratio of minutes to seconds is *always* 1:60.
    A) 60:1   B) 1:60   C) 1:360   D) 1:3600

    13. B

14. $100 \div \left(\frac{1}{10} \times \frac{1}{10}\right) = 100 \div \frac{1}{100} = 100 \times 100$.
    A) 100×100 B) 10×10 C) 10   D) 1

    14. A

15. The angles of an isosceles right triangle are 45°, 45°, and 90°.
    A) 30°   B) 45°   C) 60°   D) 90°

    15. B

16. The largest would be $99^2 = 9801$, a number with 4 digits.
    A) 4   B) 3   C) 2   D) 1

    16. A

Go on to the next page ⏩ **8**

17.	The difference between two primes > 100 is odd − odd = even. A) even B) odd C) composite D) prime	17. A
18.	Half of 1/3 of a cup of each of 5 juices will fill $(1/2) \times (1/3) \times (5) = 5/6$ of a cup. A) $\frac{1}{5}$ B) $\frac{1}{3}$ C) $\frac{5}{6}$ D) 1	18. C
19.	$1998 = 2 \times 3^3 \times 37$, so 1998 has 3 prime factors. A) 2 B) 3 C) 4 D) 5	19. B
20.	The area of a square of side 5 cm = 25 cm². The perimeter is 20 cm. The quotient is 25/20 = 1.25. A) 1.25 B) 1.2 C) 1 D) 0.8	20. A
21.	If $\overline{BD}$ is the diagonal of square *ABCD*, as shown at the right, then $m\angle ABD$ = half of 90° = 45°. A) 30° B) 45° C) 60° D) 90°	21. B
22.	(Any number) × (its reciprocal) = 1; now double to get 2. A) 0.5 B) 1 C) 1.5 D) 2	22. D
23.	Sid's snake slithers 1 km = 1000 m in 50 minutes. Its average speed is 1000/50 m/min = 20 m/min. A) 10 m/min B) 20 m/min C) 30 m/min D) 50 m/min	23. B
24.	If 19:98 = x:97, then since 98 > 97, 19 > x. A) $x < 19$ B) $x > 19$ C) $x = 18$ D) $x = 19$	24. A
25.	If $m\angle 4 + m\angle 1 = 180°$, & $m\angle 4 = (m\angle 1)/3$, then $m\angle 4 = m\angle 2 = 45°$, and $m\angle 1 = m\angle 3 = 135°$. A) 135° B) 120° C) 60° D) 45°	25. A
26.	If the 14th day is a Wed., then the 28-31st days fall on Wed.-Sat., and the 1st falls on Thurs.-Sun. A) Friday B) Saturday C) Sunday D) Monday	26. D
27.	$1 \div \frac{1}{2} + 2 \div \frac{1}{4} + 4 \div \frac{1}{8} = 2 + 8 + 32 = 42$. A) 1.5 B) 6 C) 8 D) 42	27. D
28.	Since 30% of the seats are used by 75 people, 10% are used by 25 people, and 70% are used by 7×25 = 175 people. A) 150 B) 175 C) 225 D) 250	28. B
29.	Each year, she gets 6 issues of the bimonthly and 2 issues of the semiannual magazine. Altogether, she gets 8 issues each year. A) 7 B) 8 C) 25 D) 26	29. B
30.	21 hrs = $[21/(7 \times 24)] \times 100 = 12.5\%$ of 1 week. A) 87.5 B) 21 C) 12.5 D) 8	30. C

Go on to the next page ⫸ **8**

31. The sum of 4 consecutive odds is 4 more than the sum of 4 consecutive evens, so each odd is 1 more than an even, and the least of the odds is 1 more than the least of the evens.

   A) 1   B) 2   C) 3   D) 4

31.

A

32. Reciprocal of x is > 1, so x is positive and < 1.

   A) $x < 0$ B) $x = 0$ C) $0 < x < 1$ D) $x > 1$

The **FAX**..
The Whole **FAX**...
and nothing
but the **FAX**...

32.

C

33. Vinnie, born in March of a leap year on a Tuesday, missed the extra day in his year of birth. Thus: 1 on a Wednesday, 2 on a Thursday, 3 on a Friday, 4 on a Sunday (leap year), 5 on a Monday, 6 on a Tuesday, 7 on a Wednesday; so Vinnie could have been 7.

   A) 6   B) 7   C) 8   D) 9

33.

B

34. Let the diameters be 2 and 4, with radii 1 and 2. The area ratio is π to $4\pi = 1{:}4$.

   A) 1:2   B) 1:3   C) 1:4   D) $1{:}\pi$

34.

C

35. The largest 3-digit multiple of 9 is $999 = 3 \times 3 \times 3 \times 37$. This number is divisible by 3, 27, and 37, but *not* by 17.

   A) 3   B) 17   C) 27   D) 37

35.

B

36. $\sqrt{1 + \frac{1}{4} + \frac{1}{9}} = \sqrt{\frac{49}{36}} = \frac{7}{6} = 1 + \frac{1}{2} + \frac{1}{3} - \frac{2}{3}$ A) $-\frac{2}{3}$ B) 0   C) $\frac{2}{3}$ D) $\frac{7}{6}$

36. A

37. A square with side-length 2 has area 4. The inscribed circle has diameter 2, radius 1, and area π. The difference is $4 - \pi$.

   A) $4\pi - 4$ B) $4 - 4\pi$ C) $4 - 2\pi$ D) $4 - \pi$

37.

D

38. $\frac{1}{2} + \frac{1}{2^2} = \frac{3}{4}$; $\frac{1}{2} + \frac{1}{2^2} + \frac{1}{2^3} = \frac{7}{8}$. By the pattern, it's D.

   A) $\frac{2^{19}}{2^{20}}$ B) $\frac{2^{19}-1}{2^{20}}$ C) $\frac{2^{19}+1}{2^{20}}$ D) $\frac{2^{20}-1}{2^{20}}$

38.

D

39. I added together all the digits of a certain whole number. If I subtract this sum from my original number, the result *could* be $2001 - (2+1) = 1998$.

   A) 1998   B) 1999   C) 2000   D) 2001

39.

A

40. The 9 numbers $10^2, 20^2, \ldots, 90^2$ are the only squares $< 10\,000$ that are divisible by 10.

   A) 9   B) 10   C) 99   D) 100

40.

A

The end of the contest ✍ **8**

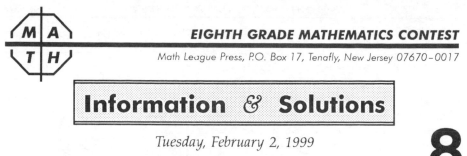

Information & Solutions

Tuesday, February 2, 1999

Contest Information

8

- **Solutions** Turn the page for detailed contest solutions (written in the question boxes) and letter answers (written in the *Answers* column to the right of each question).

- **Scores** Please remember that *this is a contest, not a test*—and there is no "passing" or "failing" score. Few students score as high as 30 points (75% correct). Students with half that, 15 points, *deserve commendation!*

- **Answers & Rating Scale** Turn to page 145 for the letter answers to each question and the rating scale for this contest.

1. $\dfrac{19-99}{99-19} = \dfrac{-80}{80} = -1.$ A) –2   B) –1   C) 0.1   D) 1

1. B

2. $12 \times 20 = 240$, and $240 \div 15 = 16$.
   A) 16   B) 18   C) 20   D) 24

2. A

3. A gumball machine has 1000 gumballs.
   At 5¢ per gumball, it costs $1000 \times 5¢ = 5000¢ = \$50.00$ to buy all the gumballs.
   A) \$20   B) \$25   C) \$50   D) \$100

3. C

4. 500% of 1 is 5, and 5 is a prime number.
   A) 1   B) 2   C) 3   D) 5

4. A

5. $(2000-1001)+(1000-1) = 999+999 = 1998 = 1999 - 1$.
   A) 1   B) –1   C) 1998   D) 2000

5. A

6. When a nonzero number is squared, the result is always positive.
   A) less than 1   B) less than 0   C) more than 1   D) more than 0

6. D

7. $\sqrt{(-1)\times(-9)\times(-9)\times(-9)} = 3\times3\times3 = 27$ A) –27   B) –9   C) 9   D) 27

7. D

8. The shaded area is 1/4 of the 20×20 square's area, or 100.
   A) 10   B) 20   C) 100   D) 200

8. C

9. $\dfrac{3\times9\times27\times81}{1\times3\times9\times27} = 81$.   A) 3   B) 9   C) 27   D) 81

9. D

10. If the circle shown is divided into 6 congruent sectors, then each central angle is 1/6 of 360°, a 60° angle.
    A) 30°   B) 45°   C) 60°   D) 100°

10. C

11. $4 = \sqrt{16} = \sqrt{19-3}$ is a whole number.
    A) 3   B) 4   C) 5   D) 6

11. A

12. For every step Cinderella's godmother takes forward, she takes 2 steps back. If she takes 30 steps forward, she takes 60 back, for a net 30 steps back from her start.
    A) 10   B) 20   C) 30   D) 60

12. C

13. $4^1+4^2+4^3+4^4 = 4\times(1+4^1+4^2+4^3) = 4\times85$.
    A) 64   B) 85   C) 100   D) 255

13. B

14. The largest positive number has the smallest positive reciprocal.
    A) $\dfrac{2}{3} = 0.66...$ B) $\dfrac{4}{3} = 1.33...$ C) $\dfrac{3}{4} = 0.75$ D) $\dfrac{3}{2} = 1.5$

14. D

Go on to the next page ⫸ **8**

15. 1 penny + 1 nickel + 1 dime = 16¢. Since 15 of these 16¢ groups are worth $2.40, there are 15 of these 3-coin groups.

    A) 60   B) 45   C) 30   D) 15

15. B

16. The ratio 12:3 = 4:1 = 32:8.

    A) 17   B) 24   C) 28   D) 32

16. D

17. We chew through 2 cm every 30 minutes. To chew through a trunk 20 cm thick takes us 10 of these half-hour segments, a total of 5 hours.

    A) 2.5   B) 5   C) 10   D) 15

17. B

18. $\dfrac{1}{2\times8\times4\times16} = \dfrac{1}{16\times4\times16} = \dfrac{1}{4^5}.$

    A) 4^2 B) 4^3 C) 4^4 D) 4^5

18. D

19. A triangle's angle-sum is 180°, and a rhombus's is 360°. Add.

    A) 360°   B) 540°   C) 720°   D) 900°

19. B

20. Since *hh:mm:ss* means *hours:minutes:seconds*, it takes 0:04:05 to reach 6 P.M. Now, add 6 hours and it is midnight.

    A) 6:04:05   B) 6:05:05   C) 7:04:05   D) 7:05:05

20. A

21. $\dfrac{7}{20} = \dfrac{35}{100} = 35\%.$ A) 5%   B) 7%   C) 35%   D) 70%

21. C

22. After discarding any factor ≈ 1, here is what remains:

    A) 0.01×0.1 B) 0.01

    C) 0.01×10 D) 0.01×100

22. D

23. Largest 2-digit prime $\times$ smallest 3-digit prime $= 97\times101 = 9797$.

    A) 1111   B) 9191   C) 9797   D) 9991

23. C

24. The ant walked 1 km = 1000 m around a square of side 1 m. It needed $1000\div4 = 250$ trips to travel $4\text{ m}\times250 = 1000\text{ m} = 1\text{ km}$.

    A) 1000   B) 250   C) 100   D) 25

24. B

25. The whole pizza's area is $12\times48\pi$ cm^2 = 576π cm^2 = $\pi\times24^2$cm^2, so $r = 24$ cm.

    A) 12 cm   B) 18 cm   C) 24 cm   D) 48 cm

25. C

26. Divide %'s by 100: $200\times20\times2\times(1/5)\times2 = 3200$.

    A) 32   B) 320   C) 3200   D) 32 000

26. C

27. 20% of $0.1 = 0.20\times0.1 = 0.02 = 2\%$.

    A) 20   B) 2   C) 0.2   D) 0.02

27. A

28. 23, a prime, is not one of the first 20 whole numbers.

    A) 23 is prime   B) $24 = 4\times6$ C) $30 = 3\times10$ D) $51 = 3\times17$

28. A

Go on to the next page ⫸ **8**

29. Sum of all 4 ∠s is 360°. Three 70° ∠s ⇒ 4th ∠ would be 150°.
    A) 70°   B) 110°   C) 147°   D) 150°

    29. D

30. The **largest** of several negative numbers is the one nearest to 0.
    A) $\sqrt{65}$ B) 8   C) 4^2 D) $\sqrt{2}$

    30. D

31. Of the last 3 choices, D is the one nearest to 1, and D ≈ 7.37. We know that A is between 0 and 1, so A is nearest to 1.
    A) 0.9^{1999} B) 1.1^{1999} C) 1.01^{1999} D) 1.001^{1999}

    31. A

32. If the area of the right triangle shown is 18, each leg = 6 = a radius, so the area of the circle is 36π.
    A) 36π B) 18π C) 90   D) 36

    32. A

33. $2^3 = 8$ and $4^3 = 64$, so we can fit 8 of the smaller in the larger.
    A) 2   B) 4   C) 8   D) 64

    33. C

34. If $a☺b$ means $a \times (a-b)$, then $2☺3 = 2 \times (2-3) = -2$.
    A) −2   B) 2   C) 3   D) 6

    34. A

35. $\frac{8}{7} \times \frac{9}{8} \times \frac{10}{9} \times \frac{11}{10} \times \frac{12}{11} \times \frac{13}{12} \times \frac{14}{13} = \frac{14}{7} = 2.$
    A) 1   B) 2   C) 7   D) 14

    35. B

36. Robinson Crusoe found a bottle that listed these primes: 101, 103, 107, 109. Since there were four primes on the list, the answer to this question is D.
    A) 1   B) 2   C) 3   D) 4

    36. D

37. $2^{1998} \times 5^{1999} \times 10^{2000} = 2^{1998} \times 5^{1998} \times 10^{2000} \times 5 = (2 \times 5)^{1998} \times 10^{2000} \times 5 = 10^{1998} \times 10^{2000} \times 5 = 10^{3998} \times 5.$
    A) 2   B) 2000   C) 3998   D) 3999

    37. C

38. Square an integer. Is 15 a factor? If so, then $15^2 = 225$ is a factor of this square, as is any factor of 225, such as 75.
    A) 30   B) 75   C) 135   D) 150

    38. B

39. Each circle has area 4π, so $r = 2$. Then, a side of the square is $6 \times r = 12$. The area of the square $= 12^2 = 144$.
    A) 144   B) 64   C) 36   D) 36π

    39. A

40. By analogy, think of a square of area 4. At first, side = 2. Tripled, it's 6; and new area is 36.
    A) 9   B) 12   C) 36   D) 48

    40. C

The end of the contest ✍ **8**

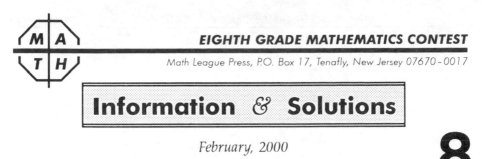
Information & Solutions

February, 2000

Contest Information

8

- **Solutions** Turn the page for detailed contest solutions (written in the question boxes) and letter answers (written in the *Answers* column to the right of each question).

- **Scores** Please remember that *this is a contest, not a test*—and there is no "passing" or "failing" score. Few students score as high as 30 points (75% correct). Students with half that, 15 points, *deserve commendation!*

- **Answers & Rating Scale** Turn to page 146 for the letter answers to each question and the rating scale for this contest.

1. The average of equally spaced numbers is the middle number. A) 5 B) 1999 C) 2000 D) 2500	1. C
2. $10 + 110 \times 0 \times 101 + 111 = 10 + 0 + 111 = 121$. A) 0 B) 120 C) 121 D) 241	2. C
3. Odd multiples of 3 are always odd, so their difference is even. A) even B) odd C) 3 D) 6	3. A
4. $\frac{1}{10} \times \left(\frac{1}{10} + \frac{2}{10} + \frac{3}{10} + \frac{4}{10} \right) = \frac{1}{10} \times \frac{10}{10} = \frac{1}{10} \times 1 = 0.1$. A) 0.1 B) 10 C) 0.01 D) 0.5	4. A
5. The drop in cost from \$2 to \$1.50 is a 50¢ decrease, and 50¢:\$2 = 50¢:200¢ = 1:4 = 25%. A) 25 B) 33⅓ C) 40 D) 50	5. A
6. The 4 days spelled with a "u" are Saturday, Sun., Tues., and Thurs. The probability is 4/7. A) $\frac{2}{7}$ B) $\frac{3}{7}$ C) $\frac{4}{7}$ D) $\frac{5}{7}$	6. C
7. $52.6\% = 0.526 < 5.2 < 21/4 = 5.25 < (5/2)^2 = 25/4 = 6.25$. A) 5.2 B) 52.6% C) $\frac{21}{4}$ D) $\left(\frac{5}{2}\right)^2$	7. D
8. $\frac{392}{5} = 78.4$, whose tens' digit is 7. A) 3 B) 4 C) 7 D) 8	8. C
9. A = 99.900 − 0.1; D > C = 99.99 + 0.1; B = 99.900 + 0.09, the closest. A) 99.8 B) 99.99 C) 100.000 D) 100.0001	9. B
10. 60 seconds = 1 min., so 60×60 seconds = 60 mins. = 1 hr. A) 36 minutes B) 600 minutes C) 6 hours D) 1 hour	10. D
11. The odd-numbered months are Jan., Mar., May, July, Sept., and Nov. Of these, Sept. and Nov. have 30 days; the rest have 31. A) 1 B) 2 C) 3 D) 4	11. B
12. My sign has perimeter 64; so each side has length $64 \div 4 = 16$. Area = $16^2 = 256$. A) 16 B) 32 C) 64 D) 256	12. D
13. Since $\sqrt{\frac{1}{4}} = \frac{1}{2}$, the sum = $4 \times \frac{1}{2} = 2$. A) 0.5 B) 1.0 C) 1.5 D) 2.0	13. D
14. Try choices, as below; OR $\frac{2}{3} \div \frac{3}{2} = \frac{4}{9} = ? \times \frac{3}{2}$; so $? = \frac{4}{9} \div \frac{3}{2} = \frac{8}{27}$. A) $\frac{8}{27} \times \frac{3}{2} = \frac{4}{9}$ B) $\frac{2}{3} \times \frac{3}{2} \neq \frac{4}{9}$ C) $\frac{3}{2} \times \frac{3}{2} \neq \frac{4}{9}$ D) $1 \times \frac{3}{2} \neq \frac{4}{9}$	14. A

Go on to the next page ⇒ **8**

15. 10 years ago, Al was as old as Bob is now, so Al is 10 years older than Bob. It will take Bob 10 years to attain Al's current age. A) 5 B) 10 C) 15 D) 20	15. B
16. Since 1 is its own reciprocal, the smallest possible (non-negative) difference is $1 - 1 = 0$. A) 0 B) 1 C) $\frac{1}{4}$ D) $\frac{1}{2}$	16. A
17. The 8 coins (average value ≈ 3¢ each) were 5 pennies, 2 nickels, and 1 dime. Total = 25¢. A) 1 nickel B) 2 dimes C) 3 nickels D) 5 pennies	17. D
18. $\frac{100+100}{199+201} = \frac{1}{2}$ and $\frac{10}{19+21} = \frac{1}{4}$. We need one more $\frac{1}{4}$, which is D. A) 1 B) $\frac{1}{9+1}$ C) $\frac{100}{19+21}$ D) $\frac{10}{19+21}$	18. D
19. half $(0.11 + $ half $0.11) = 0.5 \times (0.11 + 0.055) = 0.5 \times 0.165 = 0.0825$. A) 0.055 B) 0.0825 C) 0.11 D) 0.165	19. B
20. $(\ell,w,\mathbf{P})$ triples are: $(48,1,\mathbf{98})$, $(24,2,\mathbf{52})$, $(16,3,\mathbf{38})$, $(12,4,\mathbf{32})$, and $(8,6,\mathbf{28})$. A) 38 B) 42 C) 52 D) 98	20. B
21. The least possible sum is $0 + 1 + 2 + \ldots + 9 + 10 = 55$. A) 11 B) 45 C) 55 D) 66	21. C
22. Area $= \pi r^2 = 400\pi$, so $r^2 = 400$ and $r = 20$. Circumference $= 2\pi r = 40\pi$. A) 20π B) 40π C) 200π D) 400π	22. B
23. 5% of 5% = 10% of (half of 5%) = 10% of 2.5%. A) 0.025% B) 0.25% C) 2.5% D) 10%	23. C
24. (A prime) × (its reciprocal) = 1, and 1 is odd but not prime. A) prime B) 0 C) even D) odd	24. D
25. (A prime) ÷ (its reciprocal) = (the prime)^2. Examples: 2^2 or 3^2. A) prime B) 9 C) even D) odd	25. A
26. ⅓ perimeter = 12 < longest side < 18 = ½ perimeter. A) 11 cm B) 17 cm C) 18 cm D) 19 cm	26. B
27. The negative square root of 4 is –2, and $(-2)^3 = -8$. A) –8 B) 8 C) –6 D) –64	27. A
28. I said "no stone shall go unturned" in my search for different whole-number factors of 36. The 9 factors of 36 are 1, 2, 3, 4, 6, 9, 12, 18, and 36. A) 6 B) 8 C) 9 D) 18	28. C

Go on to the next page ▪▪▪➡ **8**

29. For the least AD, arrange A, B, C, and D as shown. Then, $AD = 9 - (3 + 2) = 4$.
    A) 2   B) 4   C) 5   D) 7

    4   2   3
    A   D C   B

    29.
    B

30. $5 \blacklozenge 12 = 5^2 + 12^2$, so $\sqrt{5 \blacklozenge 12} = \sqrt{169} = 13$.
    A) 13   B) 17   C) 60   D) 1691

    30.
    A

31. Since $1\,356\,724$ is even, both 1 and 2 are factors of $1\,356\,724$. Their positive difference is $2 - 1 = 1$.
    A) 1   B) 2   C) 3   D) 6

    31.
    A

32. $1 \div 5^{-50} = 1 \div \dfrac{1}{5^{50}} = 1 \times \dfrac{5^{50}}{1} = 5^{50}$.
    A) 5^{50} B) 5^{-50} C) $-\dfrac{1}{5^{50}}$ D) -5^{50}

    32.
    A

33. The reciprocal of 1 is 1, and the reciprocal of -1 is -1.
    A) none   B) one   C) two   D) three

    33.
    C

34. In an equilateral triangle, all three sides have equal lengths. Thus, in equilateral triangle ABC, $AB = AC$ **and** $AB = BC$.
    A) a square   B) an isosceles triangle
    C) a rectangle   D) an equilateral triangle

    34.
    D

35. In each case, fold all 4 edges of the square that is surrounded by 4 others. The shaded faces in C won't meet at a common vertex.
    A)   B)   C)   D)

    35.
    C

36. Semicircle's area $= \frac{1}{2}\pi r^2 = 2\pi$. Circle's area $= 2\pi = \pi r^2$; so $r^2 = 2$.
    A) $\dfrac{1}{4}$ B) 1   C) $\sqrt{2}$ D) 2

    36.
    C

37. The average of 2^{1999} and 2^{2001} is
    $(2^{1999} \div 2) + (2^{2001} \div 2) = 2^{1998} + 2^{2000} =$
    $2^{1998} \times (1 + 2^2) = 2^{1998} \times 5$.
    A) 5×2^{1998} B) 3×2^{1998} C) 3×2^{1999} D) 2^{2000}

    37.
    A

38. If $x^2 < x$, then $0 < x < 1$; so $\dfrac{1}{x} > 1$.
    A) 0.1   B) 0.5   C) 1   D) 2

    38.
    D

39. $1000^{100} \div (1\ \text{googol}) = (10^3)^{100} \div 10^{100} = 10^{300} \div 10^{100} = 10^{200}$.
    A) 10 googols   B) 100 googols C) (1 googol)^2 D) 2 googols

    39.
    C

40. "First $(n+5)$" pos. integers involves 5 more integers than "first n" pos. integers. Since the sum of these 5 is $240 = 46 + 47 + 48 + 49 + 50$, $n = 45$, and 45 is itself the sum of the first 9 positive integers.
    A) 9   B) 10   C) 45   D) 55

    40.
    A

The end of the contest 👉 **8**

Information & Solutions

February 20 or 27, 2001

Contest Information

- **Solutions** Turn the page for detailed contest solutions (written in the question boxes) and letter answers (written in the *Answers* column to the right of each question).

- **Scores** Please remember that *this is a contest, not a test*—and there is no "passing" or "failing" score. Few students score as high as 30 points (75% correct). Students with half that, 15 points, *deserve commendation!*

- **Answers & Rating Scale** Turn to page 147 for the letter answers to each question and the rating scale for this contest.

1. Take the inverse. Use subtraction: $10\,100 - 1010 = 9090$.
   A) 9090   B) 9900   C) 9990   D) 19 090

   1. A

2. $1+2+3+4 = 10$, so $(1+2+3+4) \times 1234 = 10 \times 1234 = 12\,340$.
   A) 10   B) 100   C) 1010   D) 1234

   2. D

3. 90 secs = 1.5 mins. It takes 40 blocks of 1.5 mins. each to make 1 hour. I earn $\$100 \div 40 = \$2.50$.
   A) 40¢   B) \$1.11   C) \$2.50   D) \$4.00

   3. C

4. 1000 grams = 1 kg, so 500 grams = 1/2 kg.
   A) 0.5   B) 2   C) 20   D) 500

   4. A

5. Choice B is closest in value to 2.5.
   A) $\sqrt{4} = 2$ B) $\sqrt{8} \approx 2.8$ C) $(0.5)^2 = 0.25$ D) $\frac{2}{5} = 0.4$

   5. B

6. The average is the middle number, 15.
   A) 15   B) 15.5   C) 16   D) 16.5

   6. A

7. The number that is 20 less than 5 is –15. Add 10: $-15+10 = -5$.
   A) –15   B) –5   C) 5   D) 25

   7. B

8. Since $0^2 = 0$ and $1^2 = 1$, there are two such whole numbers.
   A) one   B) two   C) three   D) four

   8. B

9. $1 \div (\frac{1}{2} + \frac{3}{4}) = 1 \div (\frac{2}{4} + \frac{3}{4}) = 1 \div \frac{5}{4} = 1 \times \frac{4}{5} = \frac{4}{5} = 0.8$.
   A) 0.8   B) 1   C) 1.2   D) $\frac{5}{4}$

   9. A

10. The sum of the measures of the angles of a triangle is 180°, so the average of the measures of the angles is $180° \div 3 = 60°$.
    A) 36°   B) 45°   C) 60°   D) 90°

    10. C

11. Any year with Jan. 1 a Mon., has 53 Mondays. Polly could sing 27 times.
    A) 26   B) 27   C) 104   D) 105

    11. B

12. $(4 \times 16)+(2 \times 16) = 16 \times (4+2) = 16 \times (2+4)$.
    A) 4^2 B) 4^4 C) 4^6 D) 6^6

    12. A

13. $(\$8.00+\$0.08+\$8.80) \div 8 = \$16.88 \div 8 = \$2.11$.
    A) \$2.11   B) \$2.18   C) \$2.20   D) \$3.10

    13. A

14. $7 \div 3 = 2.333 \ldots$, so the #s in D are unequal.
    A) $\frac{201}{2001}, \frac{67}{667}$ B) $0.0625, \frac{1}{16}$ C) $\frac{105}{30}, \frac{7}{2}$ D) $2.1, \frac{7}{3}$

    14. D

15. The dif. between $\frac{1}{3}$ & $\frac{1}{4}$ is $\frac{1}{3} - \frac{1}{4} = \frac{1}{12}$, so my number is $12 \times 4 = 48$.
    A) 24   B) 27   C) 36   D) 48

    15. D

Go on to the next page ⮕ **8**

16. Try an example. If Sara began with $9, she spent $3 and had $6 left. She lent $2 to her sister. She had $4 = (4/9)($9) left.

    A) $\frac{1}{3}$ B) $\frac{2}{3}$ C) $\frac{4}{9}$ D) $\frac{5}{9}$

   16. C

17. $99 \times 99 \times 99 = 99^3 = (9 \times 11)^3 = 9^3 \times 11^3 = 729 \times 11^3 = 27^2 \times 11^3$.

    A) 99×11^2 B) $27^2 \times 11^3$ C) 9×11^3 D) $3^3 \times 11^3$

   17. B

18. With 8 grades of 30 students, there are 240 students. With 24 teachers, the teacher-to-student ratio is 24:240 = 1:10.

    A) 1:10   B) 1:3   C) 4:5   D) 10:24

   18. A

19. 5% is 1/20, so 0.1% is 5% of 20(0.1%) = 2%.

    A) 2   B) 0.02   C) 0.5   D) 0.05

   19. A

20. $\frac{7+7+7}{14+21+28} = \frac{21}{63} = \frac{1}{3} = \frac{1}{2} + \frac{1}{3} + \left(-\frac{1}{2}\right)$.

    A) 0   B) $\frac{1}{4}$ C) $\frac{1}{2}$ D) $-\frac{1}{2}$

   20. D

21. There can't be more than 10 coins, since 10 nickels are worth 50¢.

    A) 4   B) 7   C) 9   D) 11

   21. D

22. Since 1 is not prime, the primes must be 2 and 3. Their sum is 5.

    A) 1   B) 3   C) 4   D) 5

   22. D

23. Any 2 adjacent ∠s are supplementary, so the sum of all 4 is 360°.

    A) 90°   B) 180°   C) 270°   D) 360°

   23. D

24. Try an example: $(1/2) \times (-1/2) = (-1/4)$, which is negative.

    A) whole   B) zero   C) positive   D) negative

   24. D

25. Circumference $= 2\pi r$, so $36 = 2\pi r$ and $r = 36 \div 2\pi = 18/\pi$.

    A) $\frac{6}{\pi}$ B) $\frac{9}{\pi}$ C) $\frac{18}{\pi}$ D) 4π

   25. C

26. If 30% of last June's rain fell in the first 15 days, 70% fell in the last 15 days. That's an average of 70% ÷ (15 days) = 4 2/3% per day.

    A) 2%   B) $2\frac{1}{3}\%$ C) $3\frac{1}{3}\%$ D) $4\frac{2}{3}\%$

   26. D

27. If one of the numbers changes from 36 to 63, then the new sum is 63−36 = 27 more than the old sum. Finally, 1234+27 = 1261.

    A) 1198   B) 1207   C) 1261   D) 1270

   27. C

28. It cost $40 for 300 rides. Dividing by 40, it cost $1 for 7.5 rides. Multiplying by 3, for $3 I'll average 3×7.5 rides = 22.5 rides.

    A) 13.3   B) 15   C) 22.5   D) 39

   28. C

29. LCM of $2^2 \times 5$, $2 \times 3 \times 5$, $2^3 \times 5$, 2×5^2, & $2^2 \times 3 \times 5$ is $2^3 \times 3 \times 5^2 = 600$.

    A) 300   B) 600   C) 1800   D) 3600

   29. B

Go on to the next page ⫸ **8**

30. Circle's area $= 64\pi$, so square's area $= 16\pi$. Its side $= \sqrt{16\pi}$.

    A) $4\sqrt{\pi}$ B) 4π C) 2   D) 16π

    30. A

31. There are 3 pairs of opposite faces: (top, bottom), (right, left), and (front, back). Use 3 different colors, one for each pair.

    A) 2   B) 3   C) 4   D) 6

    31. B

32. Patterns: $101^2 = 10\,201$; $10\,101^2 = 102\,030\,201$.

    A) 5   B) 6   C) 9   D) 18

    32. C

33. The sum of the 3 primes is 55, so all 3 are odd. The product of 3 odd numbers is always odd.

    A) 705   B) 1334   C) 1505   D) 2001

    33. B

34. Jan doesn't satisfy condition I, so Jan might or might not do well.
    I. All 13- and 14-year-old students do well on this contest.
    II. Jan is 11 years old.

    A) Jan might do well.   B) Jan will do well.
    C) Jan will not do well.   D) Jan will not take the contest.

    34. A

35. The square root of 100 is 10, but the square root of $10 \approx 3.1623$.

    A) 0   B) 1   C) 16   D) 100

    35. D

36. Use multiples of 3. Start small: $216 = 3 \times 72 = 6 \times 36 = 9 \times 24 = 12 \times 18$. Since no other multiples of 3 work, these are the only four possibilities.

    A) two B) three C) four D) six

    36. C

37. Use (0,20), (1,19), and (7,13) to get A, B, and D respectively.

    A) 0   B) 19   C) 25   D) 91

    37. C

38. The difference (2001 − choice) must be divisible by 7. D works.

    A) 1381   B) 1380   C) 1379   D) 1378

    38. D

39. As shown in the diagram at the right, $\overline{DE}$ has the same length as $\overline{BC}$.

    A D C E B

    A) $\overline{AD}$ B) $\overline{BC}$ C) $\overline{CD}$ D) $\overline{CE}$

    39. B

40. Largest odd factor is $3^{2001} \times 7^{2001} = 21^{2001}$. Its ones' digit is a 1.

    A) 1   B) 3   C) 7   D) 9

    40. A

The end of the contest ✍ **8**

114

Algebra Course 1 Solutions

1996-1997 through 2000-2001

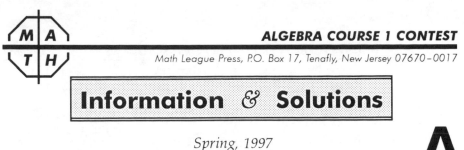

Information & Solutions

Spring, 1997

Contest Information

A

- **Solutions** Turn the page for detailed contest solutions (written in the question boxes) and letter answers (written in the *Answers* column to the right of each question).

- **Scores** Please remember that *this is a contest, not a test*—and there is no "passing" or "failing" score. Few students score as high as 30 points (75% correct). Students with half that, 15 points, *deserve commendation!*

- **Answers & Rating Scale** Turn to page 148 for the letter answers to each question and the rating scale for this contest.

1. $(1997)^{1997} + (-1997)^{1997} = (1997)^{1997} - (1997)^{1997} = 0.$
   A) 0   B) 1997^{3994} C) 2×1997^{1997} D) 3994^{1997}

   1. A

2. n (quarters) = n (5 nickels) = $5n$ nickels.
   A) $\frac{n}{5}$ B) $\frac{n}{25}$ C) $5n$ D) $25n$

   2. C

3. $\frac{-a}{-b} = \frac{a}{b} = \frac{1}{2}$ A) $\frac{-1}{2}$ B) $\frac{1}{-2}$ C) $-\frac{1}{2}$ D) $\frac{1}{2}$

   3. D

4. If $x-1 = 10^3$, then $x^2-2x+1 = (x-1)^2 = (10^3)^2 = 10^6$.
   A) 10^6+1 B) 10^6 C) 10^9+1 D) 10^9

   4. B

5. $x^2+1 = $ (real #)$^2+1 = $ (non-negative #)$+1 > 0.$
   A) none   B) 1   C) 2   D) 8

   5. A

6. $\left(\frac{1}{a}\right)^{-1} = (a^{-1})^{-1} = a^1 = \frac{2}{3}.$ A) $-\frac{3}{2}$ B) $-\frac{2}{3}$ C) $\frac{2}{3}$ D) $\frac{3}{2}$

   6. C

7. $29x^5-28x^5-23x^4+24x^4+17x^3-16x^3-13x^2+14x^2+11x-10x = $ B.
   A) $x^5 - x^4 - x^3 - x^2 + x$ B) $x^5 + x^4 + x^3 + x^2 + x$
   C) $x^5 - x^4 + x^3 - x^2 + x$ D) 5

   7. B

8. $x^2 - y^2 = (x + y)(x - y) = (10)(8) = 80.$
   A) 2   B) 4   C) 36   D) 80

   8. D

9. If there were 6 girls, half their # = 3 = ⅓ the # of boys; so the # of boys = 9, and the ratio is 6:9 = 2:3.
   A) 2:3   B) 3:2   C) 3:4   D) 4:3

   9. A

10. $x\%$ of $x = \left(\frac{x}{100}\right)(x) = \frac{x^2}{100}.$
    A) x^2 B) $\frac{x}{100}$ C) $\frac{x^2}{100}$ D) $100x^2$

    10. C

11. $|a| = |-a|$, so $|x-1| = |-(x-1)| = |-x+1| = |1-x|.$
    A) $x - 1$ B) $1 - x$ C) $|x + 1|$ D) $|1 - x|$

    11. D

12. Substituting (1996,1997) into each equation, only D is satisfied.
    A) $\frac{x}{1996} + \frac{y}{1996} = 2$ B) $\frac{x}{1997} + \frac{y}{1997} = 2$
    C) $\frac{x}{1997} + \frac{y}{1996} = 2$ D) $\frac{x}{1996} + \frac{y}{1997} = 2$

    12. D

Go on to the next page ⟹ **A**

118

13. $A=\sqrt{x^2}\sqrt{x^{1995}}$; $B=\sqrt{x^{222}}\sqrt{x^{1775}}$; $C=\sqrt{x^{888}}\sqrt{x^{1109}}$; $D=\sqrt{x^{1578}}\sqrt{x^{1417}}$.

   A) $x\sqrt{x^{1995}}$ B) $x^{111}\sqrt{x^{1775}}$ C) $x^{444}\sqrt{x^{1109}}$ D) $x^{789}\sqrt{x^{417}}$

13.

D

14. Since parallel lines cannot have slopes with opposite signs, they cannot have slopes whose product is negative.

   A) π B) $\sqrt{17}$ C) –1   D) 0

BIG BANG RESEARCH PROJECT

14.

C

15. The squares differ by 0, so the integers *could* be opposites; and $(30)(-30) = -900$.

   A) 200   B) 500   C) –800   D) –900

15.

D

16. Chris is c years old and Pat is p years old. The difference between their ages today is $c - p$. The difference between their ages 5 years ago is the same as the difference between their ages today.

   A) $c - p$ B) $c - p - 5$ C) $c - (p-5)$ D) $c - p - 10$

16.

A

17. $\dfrac{(x-1)^{10}}{(1-x)^5} = \dfrac{(1-x)^{10}}{(1-x)^5} = (1-x)^5$. **OR**, when $x = 2$, only C equals –1.   A) $(x-1)^2$ B) $(1-x)^2$ C) $(1-x)^5$ D) $(x-1)^5$

17.

C

18. $\dfrac{x+y}{\frac{1}{x}+\frac{1}{y}} = \dfrac{x+y}{\frac{x+y}{xy}} = \dfrac{\frac{1}{xy}}{\frac{1}{xy}} = xy$ A) $x+y$ B) xy C) $\frac{1}{xy}$ D) $\frac{1}{x+y}$

18.

B

19. If a positive number satisfies $4 \le x^2 \le 81$, its opposite does also; so h, the sum of all these solutions, is 0. It's now 4 P.M., so that's when my raft leaves.

   A) 4 P.M.   B) 2 P.M.   C) noon   D) 8 A.M.

HAWAII

19.

A

20. The roots of $|x| = 2$ are ±2, just as in B.

   A) $x^2+4x+4 = 0$ B) $x^2-4 = 0$
   C) $x^2-4x+4 = 0$ D) $x^2+4 = 0$

20.

B

21. The sum of the roots is $1-2+3-4+ \ldots +97-98+99-100 =$ $(1-2)+(3-4)+ \ldots +(97-98)+(99-100) = -1\times50 = -50$.

   A) –50   B) 50   C) –100   D) 100

21.

A

22. Two vertices of a square are (0,0) and (0,–4). The three possible squares are shown. The diagonals could intersect at (2,–2), (0,–2) or (–2,–2).

   A) (2,–2)   B) (–2,2)   C) (0,–2)   D) (–2,–2)

22.

B

Go on to the next page ⦀➡ **A**

119

1996-97 ALGEBRA COURSE 1 CONTEST SOLUTIONS

23. Use the choices to replace U and V. Choice A works, since
    $$\frac{-1/2}{x+2} + \frac{1/2}{x-2} = \frac{2}{x^2-4}.$$ (You could let $x = 0$ before substituting.)

    A) $\left(-\frac{1}{2}, \frac{1}{2}\right)$ B) $\left(-\frac{1}{2}, -\frac{1}{2}\right)$ C) $\left(\frac{1}{2}, -\frac{1}{2}\right)$ D) $\left(\frac{1}{2}, \frac{1}{2}\right)$

23. **A**

24. Since $p^2 = x$, it follows that $(4s)^2 = x$, or $16s^2 = x$.
    Since area $= s^2$, area $= \frac{x}{16}$

    A) x B) $\frac{x}{4}$ C) $\frac{x^2}{4}$ D) $\frac{x}{16}$

24. **D**

25. $75 = 3 \times 25$. A square can be factored into square divisors, so answer is $9 \times 25 = 225$.

    A) 375   B) 300   C) 225   D) 150

25. **C**

26. If each side is s, and the sum of the cubes of two sides equals the square of the third side, then $s^3 + s^3 = s^2$. Solving, $2s^3 = s^2$, so $2s = 1$, $s = 0.5$, and the perimeter $= 3s = 3(0.5) = 1.5$.

    A) 0.5   B) 1.5   C) 4.5   D) 6

26. **B**

27. A should say $\sqrt{a^2} = |a|$. C says $a^2 - 2a + 1 = (a-1)^2 \geq 0$. That's true!

    A) $\sqrt{a^2} = a$ B) $a^2 \geq a$ C) $a^2 + 1 \geq 2a$ D) $a \geq \frac{1}{a}$

27. **C**

28. If $dr = 100\pi$, then $(2r)(r) = 100\pi$, or $r^2 = 50\pi$. Finally, since the area of the circle is πr^2, area $= (\pi)(50\pi) = 50\pi^2$.

    A) $50\pi^2$ B) 50π C) $100\pi^2$ D) 100π

28. **A**

29. n is a square, so $\sqrt{n}$ is integral & $(\sqrt{n}-1)^2 = n - 2\sqrt{n} + 1$ is a square.

    A) $n+1$ B) $n+2\sqrt{n}$ C) n^2+4 D) $n-2\sqrt{n}+1$

29. **D**

30. **1,2,2,3,3,3,4,4,4,4**

    Add reciprocals with the same denominator and the sum is 1. Do this for all the denominators from 1 through 99 and the sum of all these reciprocals will be 99. Now, add the next 25 reciprocals of 100 to get a sum of 99.25.

    A) 98.1   B) 99.25   C) 100.5   D) 102.75

30. **B**

The end of the contest ✍ **A**

ALGEBRA COURSE 1 CONTEST

Math League Press, P.O. Box 17, Tenafly, New Jersey 07670-0017

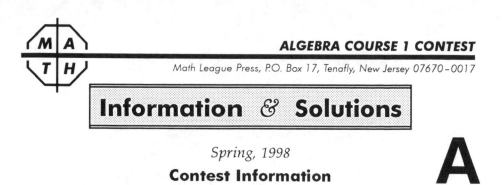

Information & Solutions

Spring, 1998

Contest Information

A

- **Solutions** Turn the page for detailed contest solutions (written in the question boxes) and letter answers (written in the *Answers* column to the right of each question).

- **Scores** Please remember that *this is a contest, not a test*—and there is no "passing" or "failing" score. Few students score as high as 24 points (80% correct). Students with half that, 12 points, *deserve commendation!*

- **Answers & Rating Scale** Turn to page 149 for the letter answers to each question and the rating scale for this contest.

1. $(10x-9x)+(8x-7x)+(6x-5x)+(4x-3x)+(2x-x) = x+\ldots+x = 5x.$

   A) x B) 5   C) $5x$ D) $10x$

   1.

   C

2. If $x + y = 0$, then $x = -y$ and $x^2 = (-y)^2 = y^2$.

   A) 0   B) y^2 C) $-y^2$ D) $-x^2$

   2.

   B

3. Since $x^2 > 25$, $x > 5$ or $x < -5$. Since $x^3 < 125$, $x < 5$. Only D satisfies all of these inequalities.

   A) 10   B) 7   C) -5 D) -144

   3.

   D

4. $\dfrac{x^2}{x-2} - \dfrac{4}{x-2} = \dfrac{x^2-4}{x-2} = 0 \Leftrightarrow x+2 = 0$, so $x = -2$.

   A) 0   B) 1   C) 2   D) 4

   4.

   B

5. Since $x-x = 0$, any product having $(x-x)$ as a factor equals 0.

   A) $x^{15} - x$ B) $x^{15} - x^5$ C) x^{10} D) 0

   5.

   D

6. $\sqrt{x} + \sqrt{4x} + \sqrt{9x} = \sqrt{x} + 2\sqrt{x} + 3\sqrt{x} = 6\sqrt{x} = \sqrt{36x}$.

   A) $\sqrt{6x}$ B) $\sqrt{14x}$ C) $\sqrt{25x}$ D) $\sqrt{36x}$

   6.

   D

7. Since $x^2 - x - 6 = (x + -3)(x + 2)$, $a = -3$.

   A) 3   B) -3 C) 8   D) -8

   7.

   B

8. If $\sqrt{x} + 2 = 0$, then $\sqrt{x} = -2$. But $\sqrt{x}$ can *never* represent a negative number, so I played for zero hours.

   A) zero   B) one   C) two   D) four

   8.

   A

9. $x\%$ of $\dfrac{1}{x} = \left(\dfrac{x}{100}\right)\left(\dfrac{1}{x}\right) = \dfrac{1}{100}$.

   A) x B) 1   C) $\dfrac{1}{x}$ D) $\dfrac{1}{100}$

   9.

   D

10. If you multiply the reciprocal of a positive number x by the opposite of x, you get $(1/x)(-x) = -1$.

    A) -1 B) 0   C) 1   D) 2

    10.

    A

11. If the sides of a quadrilateral are $\sqrt{1}$, $\sqrt{9}$, $\sqrt{9}$, and $\sqrt{8}$, then the perimeter of the quadrilateral is $1 + 3 + 3 + 2\sqrt{2} = 7 + 2\sqrt{2}$.

    A) $\sqrt{1+9+9+8}$ B) $9\sqrt{2}$ C) $7 + 2\sqrt{2}$ D) $18\sqrt{2}$

    11.

    C

Go on to the next page ▮▮▮➤ **A**

12. $17(x+3)$ is prime if $(x+3) = 1$. Hence, $x = -2$.

   A) –2   B) 0   C) 2   D) 1998

   12.

   A

13. If $P = 36x$, then $s = 9x$ and $A = (9x)^2$.

   A) $24x$ B) $24x^2$ C) $36x^2$ D) $81x^2$

   13.

   D

14. Using $a^2-b^2 = (a+b)(a-b)$, we get $(199...997)(1)$.

   A) 888 888 888 887   B) 999 999 999 997
   C) 1 888 888 888 887   D) 1 999 999 999 997

   14.

   D

15. $\sqrt{217} - \sqrt{216} < \sqrt{216} - \sqrt{215}$, so $\sqrt{217}$ is closest.

   A) $\sqrt{214}$ B) $\sqrt{215}$ C) $\sqrt{217}$ D) $\sqrt{218}$

   15.576

   15.

   C

16. $(x+1)(x+4) = (x^2+5x+4)$ & $(x+2)(x+3)$ is A.

   A) $x^2 + 5x + 6$ B) $x^2 + 4x + 3$
   C) $x^2 + 3x + 2$ D) $x^2 + 2x + 1$

   16.

   A

17. If x and y have opposite signs, then only D is true.

   A) $x^{999} = y^{999}$ B) $x^{1997} = y^{1997}$ C) $x^{1999} = y^{1999}$ D) $x^{2000} = y^{2000}$

   17.

   D

18. If $y = 0$, the x-intercept of $2x - 4y = 16$ is 8, as is the x-intercept of B.

   A) $4x - 2y = 16$ B) $2x + 4y = 16$
   C) $4y - 2x = 16$ D) $2y + 4x = 16$

   18.

   B

19. 1 isn't a prime, so $2n+1, 2n+3$, & $2n+5$ are *all* prime only if $n = 1$.

   A) $n = 0$ only B) $n = 1$ only C) $n = 0$ or 1 D) $n = 2$

   19.

   B

20. $(x-12)(x-1)$, $(x-6)(x-2)$, and $(x-4)(x-3)$ are factors for A, B, C.

   A) –13   B) –8   C) –7   D) –2

   20.

   D

21. Since $\sqrt{x^2} = |x|$, $\sqrt{x^2}$ cannot equal –1.

   A) –1   B) 0   C) 1   D) 2

   21.

   A

22. This right triangle has an altitude of length 1998 and a base of length 1998. Its area is $(1/2)(1998)(1998) = 999 \times 1998$.

   A) 1998   B) 1998^2
   C) 999×1998 D) 0.5×1998

   22.

   C

Go on to the next page ▮▮▶ **A**

23. By long division, $(x^3 - 1) \div (x - 1) = x^2 + x + 1.$

A) $x^2 - 1$ B) $x^2 + 1$

C) $x^2 + x + 1$ D) $x^2 - x + 1$

23.

C

24. $\left(a + \frac{1}{a}\right)^2 = a^2 + (a)\frac{1}{a} + (a)\frac{1}{a} + \frac{1}{a^2} = a^2 + 2 + \frac{1}{a^2}.$

A) $a^2 + \frac{1}{a^2}$ B) $a^2 + \frac{1}{a^2} + 1$ C) $a^2 + \frac{1}{a^2} + 2$ D) $a^2 + \frac{1}{a^2} + 4a$

24.

C

25. For $x \geq 0$ and $y \geq 0$, $\sqrt{xy} = \frac{x+y}{2} \Leftrightarrow xy = \frac{x^2 + 2xy + y^2}{4}$. Now

multiply by 4. You'll get $x^2 - 2xy + y^2 = 0 \Leftrightarrow (x - y)^2 = 0 \Leftrightarrow x = y.$

A) $x > y$ B) $x = y$ C) $x < y$ D) $x = 0$

25.

B

26. The sides are integers. One side is 2 more
than another. The perimeter is $x + x +$
$x+2 + x+2 = 4x+4$, which is divisible by 4.

A) 1998   B) 1999   C) 2000   D) 2002

26.

C

27. Substitute $\frac{1}{b} = \frac{a}{ab} = \frac{a}{2}$ & $\frac{1}{a} = \frac{b}{ab} = \frac{b}{2}$.

A) $a + b$ B) $\frac{a+b}{2}$ C) $\frac{a+b}{3}$ D) $\frac{a+b}{4}$

27.

B

28. There are two possible values for v: 1) Since $56! = 55! \times 56$, one
possibility is $u = 56$ and $v = 55$. But 55 is not a choice. 2) Since
$56 = 8 \times 7$, we can write $8! = 6! \times 8 \times 7$; so v could also be 6.

A) 6   B) 8   C) 56   D) 57

28.

A

29. If $-2 < x \leq 1$ and $-1 < y \leq 2$, then $-4 < xy \leq 2$.

A) 6   B) 5   C) 4   D) -3

29.

D

30. A prime is divisible by 1 and the
prime. Each new prime doubles
the number of factors (old factors plus old ones times new), so
the number of factors is a power of 2; and $1024 = 2^{10}$.

A) 1024   B) 1000   C) 676   D) 400

30.

A

The end of the contest ☜ **A**

Information & Solutions

Spring, 1999

Contest Information

A

- **Solutions** Turn the page for detailed contest solutions (written in the question boxes) and letter answers (written in the *Answers* column to the right of each question).

- **Scores** Please remember that *this is a contest, not a test*—and there is no "passing" or "failing" score. Few students score as high as 24 points (80% correct). Students with half that, 12 points, *deserve commendation!*

- **Answers & Rating Scale** Turn to page 150 for the letter answers to each question and the rating scale for this contest.

1. Add 2 to both sides of $x + 1997 = 1998$ to get $x + 1999 = 2000$. A) 1 B) 2000 C) 2001 D) 2002	1. B
2. If $x^2 = 16$, then $x = \pm 4$, so $(x-1)^2 = (4-1)^2 = 9$ or $(-4-1)^2 = 25$. A) 9 & –25 B) –9 & –25 C) –9 & 25 D) 9 & 25	2. D
3. Since $(1-x) = -1(x-1)$ and $(1+x) = (x+1)$, the product $(1-x)(1+x)$ is equal to $-1(x-1)(x+1)$, which is choice B. A) $(x-1)(1+x)$ B) $-(x-1)(x+1)$ C) $-(1-x)(1+x)$ D) $(x-1)(x+1)$	3. B
4. $1999+(1999-1999) = 1999+0 = 1999-0$. A) 0 B) 1 C) 2×1999 D) –1999	4. A
5. $(1-2) + (3-4) + (5-6) + (7-8) + (9-10) = 5\times(-1) = -5$. A) 0 B) –1 C) –5 D) –10	5. C
6. 25% less than 100% of $x = 75\%$ of $x = (75/100)x = 0.75x$. A) $0.25x$ B) $25x$ C) $0.75x$ D) $75x$	6. C
7. If $m - a + t - h = 0$, add a to both sides to get $m + t - h = a$. A) $h - m - t$ B) $m + h - t$ C) $m + t - h$ D) $m + h + t$	7. C
8. The sum must be divisible by 3; and $-1 + 0 + 1 = 0$. A) 0 B) 2 C) 4 D) 8	8. A
9. The roots of $(x-3)(x-4) = 0$ are 3 and 4. Their product is 12. A) 7 B) 12 C) –7 D) –12	9. B
10. If two numbers are reciprocals, their product is always equal to 1. A) 1 B) x^2 C) y^2 D) $\frac{1}{x^2}$	10. A
11. $(1999)+(-1997) = 2$, so their average is 1. A) –2001 B) –2000 C) –1998 D) –1997	11. D
12. For $n = 1998$ and $n = 1999$, $(\sqrt{n - 1998})(\sqrt{1999 - n}) = 0$. A) 0 B) 1998 only C) 1999 only D) both 1998 & 1999	12. D

Go on to the next page ▐▶ **A**

13. The **SUM** $0 + 0 + 0 + \ldots + (-1998) = -1998$ shows that, of the 1999 real numbers, none *must* be positive.

   A) none   B) 1   C) 1997   D) 1998

14. A **PRODUCT** cannot be negative unless an odd number of its factors are negative. Thus, at least 1 factor *must* be negative.

   A) none   B) 1   C) 1997   D) 1998

15. Dividing by 3, $51x+24y = 75$ becomes $17x+8y = 25$. Change the constant from 25 to another number to get a line parallel to $17x+8y = 25$.

   A) $17x+16y = 25$ B) $17x+8y = 50$
   C) $34x+8y = 50$ D) $34x+16y = 50$

16. $(a+\frac{1}{a})(a+\frac{1}{a}) = a(a+\frac{1}{a}) + \frac{1}{a}(a+\frac{1}{a}) = a^2+1+1+\frac{1}{a^2} = a^2+\frac{1}{a^2}+2.$

   A) a B) 0   C) 1   D) 2

17. The five consecutive evens could be $a+3, a+1, a-1, a-3$, and $a-5$.

   A) $a-9$ B) $a-7$ C) $a-5$ D) $a-2$

18. Our spring break lasted $a+b$ days, where $\sqrt{a}+\sqrt{b} = \sqrt{8}$. Since $\sqrt{8} = 2\sqrt{2} = \sqrt{2}+\sqrt{2}$, $a+b = 2+2 = 4$.

   A) 4   B) 6   C) 8   D) 16

19. If $x < 0$, $|x| > 0$; so $|x| = -x$.

   A) x^2 B) x C) $x-2x$ D) $-2x$

20. If $x < 0$, $\sqrt{x^2} > 0$; so $\sqrt{x^2} = |x| = -x$, as in the solution to #19.

   A) $-x$ B) x C) $\sqrt{-x^2}$ D) $-\sqrt{x^2}$

21. Sketch the parallelogram through $(-2,3)$, $(3,1)$, $(-1,4)$, and $(-6,6)$. The diagonal through $(-2,3)$ and $(-1,4)$ has a slope of 1.

   A) -1 B) $\frac{-2}{5}$ C) $\frac{5}{9}$ D) 1

Go on to the next page ➡ **A**

22. Call the other leg x. Since $(\sqrt{a})^2+x^2 = (\sqrt{c})^2$, $x^2 = c-a$. Taking the positive square root of each side, $x = \sqrt{c-a}$.

    A) $\sqrt{c^2-a^2}$ B) $\sqrt{c^2+a^2}$ C) $\sqrt{c-a}$ D) $c+a$

22.

C

23. Since $n > 0$, $14\,400 \le n^2 \le 16\,900 \Rightarrow$ $120 \le n \le 130$, a total of 11 integers.

    A) 2   B) 9   C) 10   D) 11

23.

D

24. Dividing by 1999, $\dfrac{1999}{x} < 1 \Leftrightarrow \dfrac{1}{x} < \dfrac{1}{1999}$.

    A) $\dfrac{1}{x} < \dfrac{1}{1999}$ B) $\dfrac{x}{1999} > 1$ C) $x > 1999$ D) $\dfrac{1998}{x} < 0$

24.

A

25. If two integers have a fixed sum, the sum of their reciprocals is smallest when the two integers are as nearly equal as possible.

    A) 996 & 1004   B) 997 & 1003   C) 998 & 1002   D) 999 & 1001

25.

D

26. $1.25 \times 10^n = 125 \times 10^{n-2} = 5^3 \times 10^{n-2}$; so $n-2$ must be a multiple of 3.

    A) 1999   B) 2000   C) 2001   D) 2002

26.

B

27. Every 5th term of $1 \times 2 \times 3 \times \ldots \times 98 \times 99 \times 100$ has 5 as a factor. Every 25th has a second 5. Adding, $[100 \div 5] + [100 \div 25] = 24$.

    A) 5^2 B) 5^{20} C) 5^{24} D) 5^{100}

27.

C

28. To find the x-intercept, set $y = 0$ and solve for x. Thus, $\dfrac{3}{2x-1} = \dfrac{1}{x-2}$, so $3x-6 = 2x-1$ and $x = 5$.

    A) 7   B) 5   C) 3   D) 2

28.

B

29. Since $a+b = 50$ and $ab = 282$, the hypotenuse $= \sqrt{a^2+b^2} = \sqrt{(a+b)^2 - 2ab} = \sqrt{2500-564} = \sqrt{1936} = 44$.

    A) 44   B) 42   C) 40   D) $25\sqrt{2}$

29.

A

30. The real root of $x^3 - 1999x^2 + x - 1999 = 0$ is 1999, so the real root of $(x-1)^3 - 1999(x-1)^2 + (x-1) - 1999 = 0$ is $x-1 = 1999$, or $x = 2000$.

    A) 1   B) 1998   C) 1999   D) 2000

30.

D

The end of the contest ✍ **A**

Information & Solutions

Spring, 2000

Contest Information

■ **Solutions** Turn the page for detailed contest solutions (written in the question boxes) and letter answers (written in the *Answers* column to the right of each question).

■ **Scores** Please remember that *this is a contest, not a test*—and there is no "passing" or "failing" score. Few students score as high as 24 points (80% correct). Students with half that, 12 points, *deserve commendation!*

■ **Answers & Rating Scale** Turn to page 151 for the letter answers to each question and the rating scale for this contest.

1. $(x-1999)(1999-x) = (2000-1999)(1999-2000) = (1)(-1) = -1$.

   A) -1 B) 0 C) 1 D) 2000

2. Since $0.5x = \frac{1}{2}x = \frac{x}{2}$, its reciprocal is $\frac{2}{x}$.

   A) $\frac{2}{x}$ B) $2x$ C) $0.5x$ D) $-0.5x$

3. Dr. Roentgen finished $-1-(-1)-(-1) =$
   $-1+1+1 = 1$ hour after noon $= 1$ P.M.

   A) 1 P.M.   B) 2 P.M.   C) 3 P.M.   D) 4 P.M.

4. $(x-3)(x+4) = x^2+x-12 = (x+3)(x-4)+(2x)$.

   A) x B) $2x$ C) $-x$ D) $-2x$

5. Since (# sides hexagon) $= 2$(# sides $\triangle$), perimeters are equal
   only if (length of each hex. side) $= \frac{1}{2}$(length of each $\triangle$ side).

   A) $0.5x$ cm   B) x cm   C) $1.5x$ cm   D) $2x$ cm

6. Squaring, $a^2-2a+1 = a^2+2a+1$. This will be true only if $a = 0$.

   A) -0.25 B) 0 C) 0.25 D) 0.5

7. There are 99 integers from 2001 to 2099, and each has 0 as
   its hundreds' digit. The product of the digits is always 0.

   A) 19   B) 20   C) 99   D) 100

8. $x/y = y/k \Rightarrow kx = y^2$, so $k = y^2/x$.

   A) $\frac{y}{x}$ B) $\frac{x}{y}$ C) $\frac{y^2}{x}$ D) $\frac{x}{y^2}$

   CHANGING ADDRESSES.

9. Sum of the first 10 pos. odd integers is
   $10^2 = 100$. My apartment # is $\sqrt{100} = 10$.

   A) 10   B) 19   C) 20   D) 50

10. $x \neq 0$, so divide both sides by x to get $x = -2000$.

    A) -2000 B) -1 C) 1 D) 2000

11. If $x > 1$, then $\frac{1}{x-1} - \frac{x}{x-1} = \frac{1-x}{x-1} = -1 = \frac{-(x+1)}{x+1} = \frac{-x-1}{x+1}$.

    A) $x + 1$ B) $x - 1$ C) $1 - x$ D) $-x - 1$

12. The product of 1999 "-1"s and one "$-\pi$" is π.

    A) -2000 B) -1 C) 0 D) π

Go on to the next page ⟩⟩⟩ **A**

13. $\frac{1}{x} + \frac{1}{x^2} = \frac{x^2}{x^3} + \frac{x}{x^3} = \frac{x + x^2}{x^3}$. Now divide by $(x + x^2)$ to get $\frac{1}{x^3}$.

A) $(x+1)^2$ B) $\frac{1}{x^3}$ C) 1   D) 2

13.

B

14. Let $y = 0$ to find x-intercepts and $x = 0$ to find y-intercepts. Only in choice B will both intercepts be 1.

A) $x - y = 1$ B) $x + y = 1$
C) $x + y = -1$ D) $y = x + 1$

14.

B

15. Given $= \left(\frac{x}{x}\right)\left(\frac{x^2}{x^2}\right)\left(\frac{x^3}{x^3}\right)\left(\frac{x^4}{x^4}\right)\left(\frac{x^5}{x^5}\right)\left(\frac{1}{x^6}\right) = \frac{1}{x^6}$.

A) 1   B) x C) $\frac{1}{x}$ D) $\frac{1}{x^6}$

15.

D

16. If $3x : 5y = 7 : 11$, then $x : 5y = 7 : 33$, and $x : y = 35 : 33$.

A) $2 : 3$ B) $21 : 55$ C) $35 : 33$ D) $55 : 21$

16.

C

17. $\frac{\left(x^{2000}\right)^{2000}}{\left(x^{2000}\right)\left(x^{2000}\right)} = \frac{x^{4\,000\,000}}{x^{4000}} = x^{3\,996\,000}$.

A) 1   B) x^{2000} C) x^{4000} D) $x^{3\,996\,000}$

17.

D

18. All integers n, $-199 \le n \le 99$, work. Be sure to include $n = 0$.

A) 100   B) 298   C) 299   D) 300

18.

C

19. Broccoli + banana = 186¢, and broccoli + orange = 162¢. Now, when you subtract you'll get banana − orange = 24¢. Thus, the banana cost 24¢ more than the orange.

A) 4¢   B) 6¢   C) 12¢   D) 24¢

19.

D

20. $(1 - x^2) \div (x + 1) = (1 - x)(1 + x) \div (x + 1) = 1 - x$.

A) $x + 1$ B) $1 - x$ C) $-x$ D) $x - 1$

20.

B

21. $m_{RS} = \frac{b+d}{-a+c} = m_{AT} \ne m_{BT} = \frac{b+d}{-c-a} \ne m_{CT} = \frac{-b+d}{c-a} \ne m_{DT} = \frac{-b+d}{-c-a}$.

A) (c, b) B) $(-c, b)$ C) $(c, -b)$ D) $(-c, -b)$

21.

A

22. $(1-2)+(3-4)+ \ldots +(1999-2000) = (-1)+(-1)+ \ldots +(-1) = -1000$.

A) -2000 B) -1000 C) -1 D) 0

22.

B

Go on to the next page ⟳ **A**

131

23. Perimeter $= 2(w+\ell) = 2w+2\ell$. Area $= \ell w$. Ratio of perimeter to area $= (2w+2\ell):\ell w = (2w/\ell w + 2\ell/\ell w):1 = (2/\ell + 2/w):1$.

   A) $(\ell+w):\ell w$ B) $\ell w:(2\ell+2w)$ C) $\left(\frac{2}{\ell} + \frac{2}{w}\right):1$ D) $2:1$

23.

C

24. Since a and b are integers, the least possible $b = a + 1$. Since x is also an integer, the least possible $a + x = b + 1 = a + 2$.

   A) -1 B) 0 C) 1 D) 2

24.

D

25. The constant term of $(2x^2+3x+2)^n$ is 2^n; and $2^n = 1024$ when $n = 10$.

   A) 4 B) 8 C) 9 D) 10

25.

D

26. To go 60 km at 50 km/hr, I need $60 \div 50 = 1.2$ hrs. To go 30 km at 30 km/hr takes $30 \div 30 = 1$ hr. This leaves 0.2 hrs for the next 30 km. My speed will need to average $30 \div 0.2 = 150$ km/hr.

   A) 70 km/hr   B) 80 km/hr   C) 100 km/hr   D) 150 km/hr

26.

D

27. Change the sign of the middle term to change each root's sign.

   A) $x^2+999999x-9999 = 0$ B) $x^2-999999x+9999 = 0$
   C) $x^2+999999x+9999 = 0$ D) $x^2-9999x-999999 = 0$

27.

A

28. Let "big" negatives have odd exponents; let the negatives closest to 0 have the even exponents: $s = -1, n = -2, h = -3, t = -4, a = -5, k = -6$. Value $= (-4)^1 + (-3)^2+(-5)^3+(-2)^4+(-6)^5+(-1)^6 = -7879$.

   A) -209 B) -2229 C) -7879 D) -7897

28.

C

29. Be sure to *add* the exponents: $(3^{1+x})(3^{1-x}) = 3^{(1+x)+(1-x)} = 3^2 = 9$.

   A) 3 B) 9 C) 3^{1-x^2} D) 9^{1-x^2}

29.

B

30. $\sqrt{1000000} = \sqrt{2^6 5^6} = a\sqrt{b} = \sqrt{a^2 b}$. The 16 possible values of a^2 are $1, 5^2, 5^4, 5^6, 2^2, 2^2 5^2, 2^2 5^4, 2^2 5^6, 2^4, 2^4 5^2, 2^4 5^4, 2^4 5^6, 2^6, 2^6 5^2, 2^6 5^4,$ and $2^6 5^6$.

   A) 2 B) 4 C) 12 D) 16

30.

D

The end of the contest ✍ **A**

Information & Solutions

Spring, 2001

Contest Information

A

- **Solutions** Turn the page for detailed contest solutions (written in the question boxes) and letter answers (written in the *Answers* column to the right of each question).

- **Scores** Please remember that *this is a contest, not a test*—and there is no "passing" or "failing" score. Few students score as high as 24 points (80% correct). Students with half that, 12 points, *deserve commendation!*

- **Answers & Rating Scale** Turn to page 152 for the letter answers to each question and the rating scale for this contest.

1. If $x = 10$, then $(x - 1)^2 + (1 - x)^2 = 9^2 + (-9)^2 = 81 + 81 = 162$. A) 162 B) 81 C) 9 D) 0	1. A
2. Since $x = -1$ and $y = 2$, the intersection point has an x-coordinate of -1 and a y-coordinate of 2. A) (1,2) B) (-1,2) C) (1,-2) D) (-1,-2)	2. B
3. Since $x + y = 0$, $x = -y$ and $x^2 = (-y)^2 = y^2$. A) 0 B) y^2 C) $-y^2$ D) $-x^2$	3. B
4. $\sqrt{1-x}$ is largest when x is smallest, so choice D is correct. A) 1 B) $\frac{1}{4} = \frac{16}{64}$ C) $\frac{5}{16} = \frac{20}{64}$ D) $\frac{15}{64}$	4. D
5. Plot points. The 4th vertex has the same x-coordinate as (4001,1) and the same y-coordinate as (2001,2001). Its coordinates are (4001,2001). A) (1,2001) B) (2001,4001) C) (4001,2001) D) (4001,4001)	5. C
6. The roots of $(x-3)(x-667) = 0$ are 3 and 667, and $3 \times 667 = 2001$. A) 770 B) -770 C) 2001 D) -2001	6. C
7. Every real number between 99 and 100 satisfies $\frac{1}{100} < \frac{1}{x} < \frac{1}{99}$. A) none B) 1 C) 2 D) more than 2	7. D
8. All prime numbers greater than 2001 are odd numbers. Since the sum of two odd numbers is an even number, choice C is even and cannot be a prime number. A) $p - 2$ B) $p + 2$ C) $p + 999$ D) $p + 9990$	8. C
9. Add a to, and subtract t from, both sides. INFORMATION A) $m-t = a-h$ B) $m+t = a-h$ C) $m-t = a+h$ D) $m+t = a+h$	9. A
10. When $x = 2001$, we get $(2001-1)^2 = 2000^2$ and $(2001+1)^2 = 2002^2$. A) -2003 B) -2001 C) -1999 D) 2001	10. D
11. $(-2001+1)^2 = (-2000)^2 = 2000^2$ & $(-2001-1)^2 = (-2002)^2 = 2002^2$. A) -2003 B) -2001 C) -1999 D) 2001	11. B

Go on to the next page ⟫ **A**

12. If $2000x - 2000 = 1000$, then $x = 1.5$ and $2001x - 2001 = 1000.5$.

    A) 1000.5   B) 1001   C) 1001.5   D) 1002

    12.

    A

13. The graph of $x = -1$ is a vertical line; $y = 2$ is a horizontal line.

    A) $x = 1$ B) $x = 0$ C) $x + y = 1$ D) $y = 2$

    13.

    D

14. The difference between 1/3 of my goal and 1/4 of my goal is 1/12 of my goal. Since 1/12 of my goal is 2.5 scoops, multiply by 12 to find that my daily goal is 30 scoops.

    A) 12   B) 24   C) 27   D) 30

    14.

    D

15. $\sqrt{(2000 - 2001)^2} = |2000 - 2001| = 1$.

    A) 2000   B) $2001 - 2000$
    C) $2000 - 2001$ D) 2001

    15.

    B

16. $1 + (x+1)(x+2) + x = 1 + (x^2 + 3x + 2) + x = x^2 + 4x + 3 = (x+1)(x+3)$.

    A) x B) $x + 2$ C) $x + 3$ D) $x + 4$

    16.

    C

17. $x^2 - 98x - 99 = (x - 99)(x + 1) = 0$, so $x = 99$ or -1, and $99 - 1 = 98$.

    A) -99 B) -98 C) 98   D) 99

    17.

    C

18. $\sqrt{n} + \sqrt{n} = \sqrt{2n} \Leftrightarrow 2\sqrt{n} = \sqrt{2n}$. Squaring, $4n = 2n$, so $n = 0$.

    A) none   B) 1   C) 2   D) more than 2

    18.

    B

19. $x^4 - 16 = (x^2 + 4)(x^2 - 4) = (x^2 + 4)(x + 2)(x - 2)$.

    A) $x^2 + 4$ B) $x^2 - 4$ C) $x - 2$ D) $(x + 2)^2$

    19.

    D

20. In a square with side s and diagonal d, $d^2 = s^2 + s^2 = 2s^2$. Thus, $400 = 4s^2 + 2d^2 = 4s^2 + 4s^2 = 8s^2$, and $50 = s^2 = $ the area of the square.

    A) 50   B) 100   C) 150   D) 200

    20.

    A

21. $x^{2000} + (-x)^{2000} = x^{2000} + x^{2000} = 2x^{2000}$.

    A) $2x^{2000}$ B) $2x^{2001}$ C) $|x|^{2001}$ D) 0

    21.

    A

22. The LCM is x^4 since $x^4 = (x^3)x = (x^2)(x^2)$.

    A) x^4 B) x^6 C) x^{12} D) x^{24}

    22.

    A

Go on to the next page ⟫ **A**

135

23. Substitute in C: $-|a| + |b| = -(-a) + (-b) = a - b$. A) $|a| + |b|$ B) $|a| - |b|$ C) $-|a| + |b|$ D) $-|a| - |b|$	23. C
24. Rewrite $2x-y = 3$ as $y = 2x-3$. Its slope is 2. Rewrite each choice in the form $y = mx+b$. Choice A is $y = 4x-3$. A) $4x-y = 3$ B) $2x-2y = 3$ C) $4x-2y = 6$ D) $2x-2y = 6$	24. A
25. If 1 million $< n^2 <$ 4 million, then $1000 < n < 2000$. The 999 possible values of n are 1001, 1002, . . . , 1999. A) 998 B) 999 C) 1000 D) 1998	25. B
26. $\sqrt{2001} \approx 44.73$, so the 45 possible integers are 0, 1, . . . , 43, 44. A) 42 B) 43 C) 44 D) 45	26. D
27. Dividing by 7, $21x+28y = 84 \Leftrightarrow 3x+4y = 12$. Doubling, $6x+8y = 24$. A) 12 B) 20 C) 24 D) 48	27. C
28. In a *pseudopythagorean* triangle, $\sqrt{a} + \sqrt{b} = \sqrt{c}$. Squaring both sides, $a+2\sqrt{ab} +b = c$, implying that $a+b < c$. But, in any triangle, $a+b > c$. Therefore, no such triangle is possible. A) none B) 1 C) 2 D) more than 2	28. A
29. If $5^x = 7$, then $5^{x+2} = (5^x)(5^2) = 7 \times 25 = 175$. A) 9 B) 14 C) 49 D) 175	29. D
30. $2x+3y = 9\,999\,999\,999\,999 \Leftrightarrow 2x = 3(3\,333\,333\,333\,333 - y)$. The right side is divisible by 3, so the left is too, and $x = 3$, a prime. A) none B) one C) two D) three	30. B

The end of the contest 🖅 **A**

136

Answer Keys &
Difficulty Ratings

• • • • • • • • • • • • • • • • •

1996-1997 through 2000-2001

ANSWERS, 1996-97 7th Grade Contest

1. B	9. B	17. A	25. B	33. B
2. A	10. C	18. B	26. D	34. D
3. D	11. A	19. B	27. B	35. D
4. C	12. D	20. C	28. A	36. B
5. B	13. C	21. A	29. D	37. A
6. C	14. D	22. C	30. C	38. A
7. A	15. B	23. D	31. A	39. D
8. D	16. C	24. D	32. C	40. B

RATE YOURSELF!!!
for the 1996-97 7th GRADE CONTEST

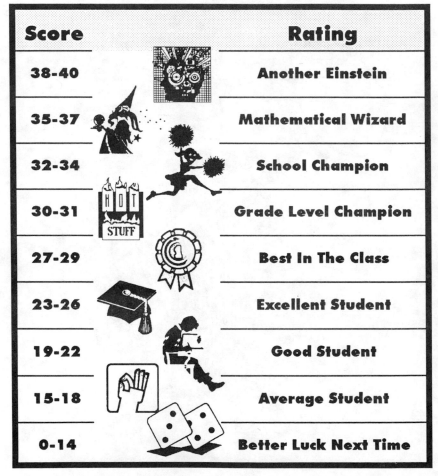

Score	Rating
38-40	Another Einstein
35-37	Mathematical Wizard
32-34	School Champion
30-31	Grade Level Champion
27-29	Best In The Class
23-26	Excellent Student
19-22	Good Student
15-18	Average Student
0-14	Better Luck Next Time

ANSWERS, 1997-98 7th Grade Contest

1. B	9. D	17. A	25. A	33. A
2. A	10. B	18. D	26. B	34. B
3. D	11. A	19. B	27. B	35. C
4. D	12. C	20. D	28. C	36. B
5. C	13. C	21. A	29. C	37. C
6. A	14. B	22. D	30. C	38. D
7. D	15. C	23. B	31. B	39. D
8. D	16. A	24. A	32. C	40. A

RATE YOURSELF!!!
for the 1997-98 7th GRADE CONTEST

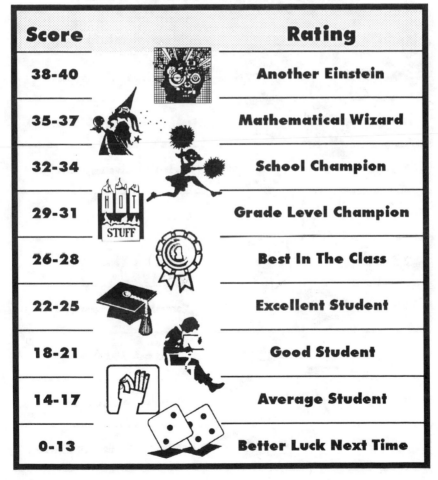

Score	Rating
38-40	Another Einstein
35-37	Mathematical Wizard
32-34	School Champion
29-31	Grade Level Champion
26-28	Best In The Class
22-25	Excellent Student
18-21	Good Student
14-17	Average Student
0-13	Better Luck Next Time

ANSWERS, 1998-99 7th Grade Contest

1. C	9. B	17. C	25. C	33. B
2. A	10. C	18. A	26. D	34. A
3. A	11. A	19. B	27. B	35. D
4. C	12. D	20. A	28. B	36. C
5. A	13. C	21. D	29. D	37. C
6. B	14. B	22. C	30. C	38. A
7. D	15. B	23. D	31. A	39. B
8. D	16. D	24. A	32. C	40. B

RATE YOURSELF!!!
for the 1998-99 7th GRADE CONTEST

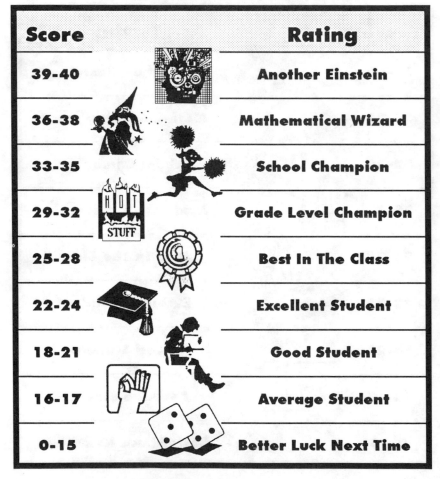

Score	Rating
39-40	Another Einstein
36-38	Mathematical Wizard
33-35	School Champion
29-32	Grade Level Champion
25-28	Best In The Class
22-24	Excellent Student
18-21	Good Student
16-17	Average Student
0-15	Better Luck Next Time

ANSWERS, 1999-00 7th Grade Contest

1. C	9. B	17. B	25. C	33. B
2. A	10. C	18. A	26. B	34. A
3. A	11. A	19. C	27. C	35. B
4. D	12. D	20. B	28. D	36. B
5. A	13. C	21. D	29. C	37. C
6. B	14. B	22. C	30. A	38. C
7. A	15. D	23. A	31. D	39. B
8. C	16. D	24. A	32. D	40. A

RATE YOURSELF!!!
for the 1999-00 7th GRADE CONTEST

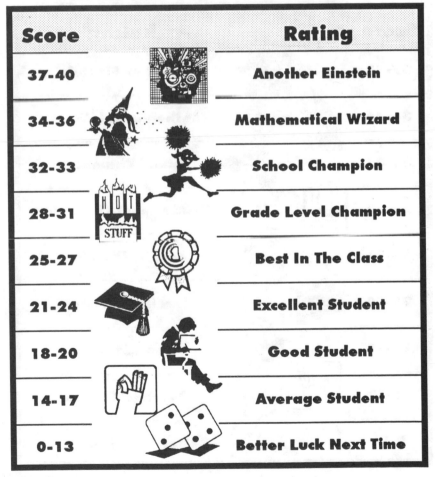

Score	Rating
37-40	Another Einstein
34-36	Mathematical Wizard
32-33	School Champion
28-31	Grade Level Champion
25-27	Best In The Class
21-24	Excellent Student
18-20	Good Student
14-17	Average Student
0-13	Better Luck Next Time

ANSWERS, 2000-01 7th Grade Contest

1. D	9. C	17. C	25. B	33. D
2. A	10. B	18. A	26. A	34. B
3. C	11. A	19. D	27. C	35. A
4. C	12. A	20. B	28. C	36. D
5. D	13. A	21. B	29. B	37. C
6. D	14. D	22. D	30. B	38. B
7. C	15. C	23. B	31. A	39. C
8. B	16. A	24. D	32. C	40. B

RATE YOURSELF!!!
for the 2000-01 7th GRADE CONTEST

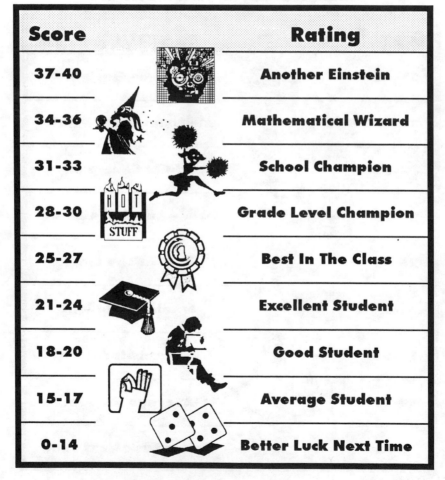

Score	Rating
37-40	Another Einstein
34-36	Mathematical Wizard
31-33	School Champion
28-30	Grade Level Champion
25-27	Best In The Class
21-24	Excellent Student
18-20	Good Student
15-17	Average Student
0-14	Better Luck Next Time

ANSWERS, 1996-97 8th Grade Contest

1. C	9. A	17. B	25. B	33. B
2. A	10. B	18. A	26. A	34. C
3. B	11. B	19. D	27. A	35. A
4. A	12. D	20. C	28. D	36. A
5. C	13. D	21. A	29. D	37. D
6. C	14. B	22. C	30. C	38. D
7. A	15. C	23. B	31. D	39. B
8. D	16. D	24. C	32. C	40. B

RATE YOURSELF!!!
for the 1996-97 8th GRADE CONTEST

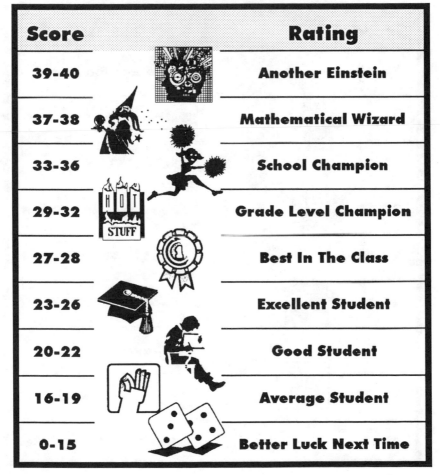

Score	Rating
39-40	Another Einstein
37-38	Mathematical Wizard
33-36	School Champion
29-32	Grade Level Champion
27-28	Best In The Class
23-26	Excellent Student
20-22	Good Student
16-19	Average Student
0-15	Better Luck Next Time

ANSWERS, 1997-98 8th Grade Contest

1. C	9. C	17. A	25. A	33. B
2. C	10. C	18. C	26. D	34. C
3. D	11. A	19. B	27. D	35. B
4. A	12. C	20. A	28. B	36. A
5. D	13. B	21. B	29. B	37. D
6. D	14. A	22. D	30. C	38. D
7. B	15. B	23. B	31. A	39. A
8. B	16. A	24. A	32. C	40. A

RATE YOURSELF!!!
for the 1997-98 8th GRADE CONTEST

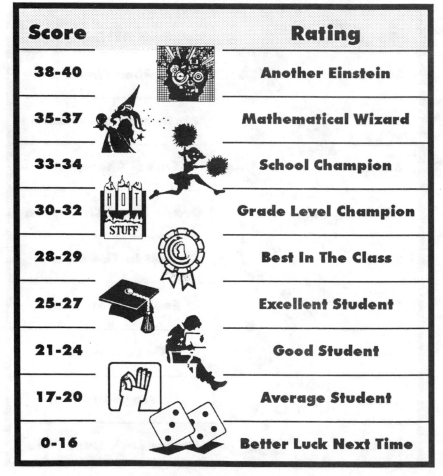

Score	Rating
38-40	Another Einstein
35-37	Mathematical Wizard
33-34	School Champion
30-32	Grade Level Champion
28-29	Best In The Class
25-27	Excellent Student
21-24	Good Student
17-20	Average Student
0-16	Better Luck Next Time

ANSWERS, 1998-99 8th Grade Contest

1. B	9. D	17. B	25. C	33. C
2. A	10. C	18. D	26. C	34. A
3. C	11. A	19. B	27. A	35. B
4. A	12. C	20. A	28. A	36. D
5. A	13. B	21. C	29. D	37. C
6. D	14. D	22. D	30. D	38. B
7. D	15. B	23. C	31. A	39. A
8. C	16. D	24. B	32. A	40. C

RATE YOURSELF!!!
for the 1998-99 8th GRADE CONTEST

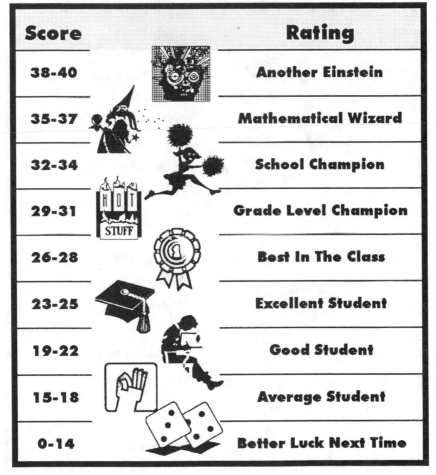

Score	Rating
38-40	Another Einstein
35-37	Mathematical Wizard
32-34	School Champion
29-31	Grade Level Champion
26-28	Best In The Class
23-25	Excellent Student
19-22	Good Student
15-18	Average Student
0-14	Better Luck Next Time

ANSWERS, 1999-00 8th Grade Contest

1. C	9. B	17. D	25. A	33. C
2. C	10. D	18. D	26. B	34. D
3. A	11. B	19. B	27. A	35. C
4. A	12. D	20. B	28. C	36. C
5. A	13. D	21. C	29. B	37. A
6. C	14. A	22. B	30. A	38. D
7. D	15. B	23. C	31. A	39. C
8. C	16. A	24. D	32. A	40. A

RATE YOURSELF!!!
for the 1999-00 8th GRADE CONTEST

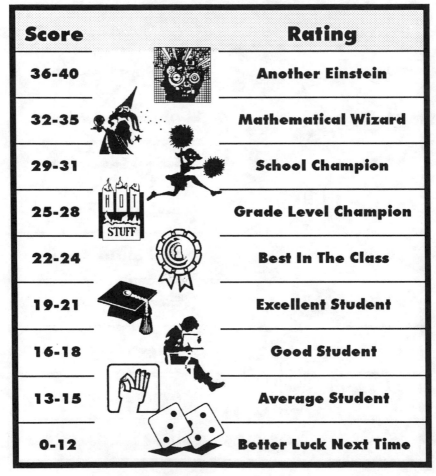

Score	Rating
36-40	Another Einstein
32-35	Mathematical Wizard
29-31	School Champion
25-28	Grade Level Champion
22-24	Best In The Class
19-21	Excellent Student
16-18	Good Student
13-15	Average Student
0-12	Better Luck Next Time

ANSWERS, 2000-01 8th Grade Contest

1. A	9. A	17. B	25. C	33. B
2. D	10. C	18. A	26. D	34. A
3. C	11. B	19. A	27. C	35. D
4. A	12. A	20. D	28. C	36. C
5. B	13. A	21. D	29. B	37. C
6. A	14. D	22. D	30. A	38. D
7. B	15. D	23. D	31. B	39. B
8. B	16. C	24. D	32. C	40. A

RATE YOURSELF!!!
for the 2000-01 8th GRADE CONTEST

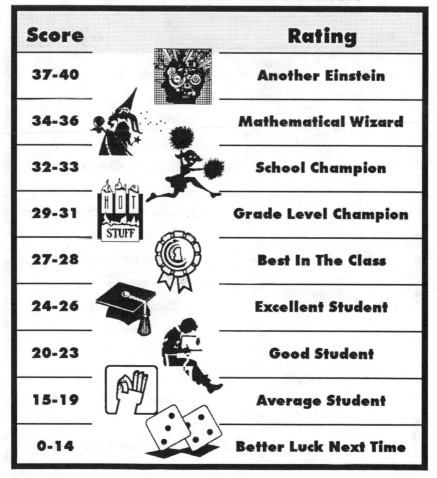

Score	Rating
37-40	Another Einstein
34-36	Mathematical Wizard
32-33	School Champion
29-31	Grade Level Champion
27-28	Best In The Class
24-26	Excellent Student
20-23	Good Student
15-19	Average Student
0-14	Better Luck Next Time

ANSWERS, 1996-97 Algebra Course 1 Contest

1. A	7. B	13. D	19. A	25. C
2. C	8. D	14. C	20. B	26. B
3. D	9. A	15. D	21. A	27. C
4. B	10. C	16. A	22. B	28. A
5. A	11. D	17. C	23. A	29. D
6. C	12. D	18. B	24. D	30. B

RATE YOURSELF!!!
for the 1996-97 ALGEBRA COURSE 1 CONTEST

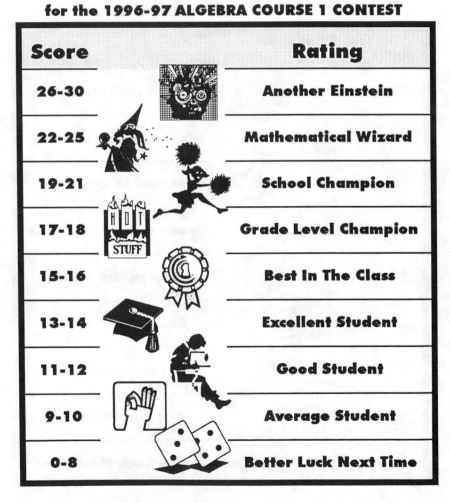

Score	Rating
26-30	Another Einstein
22-25	Mathematical Wizard
19-21	School Champion
17-18	Grade Level Champion
15-16	Best In The Class
13-14	Excellent Student
11-12	Good Student
9-10	Average Student
0-8	Better Luck Next Time

ANSWERS, 1997-98 Algebra Course 1 Contest

1. C	7. B	13. D	19. B	25. B
2. B	8. A	14. D	20. D	26. C
3. D	9. D	15. C	21. A	27. B
4. B	10. A	16. A	22. C	28. A
5. D	11. C	17. D	23. C	29. D
6. D	12. A	18. B	24. C	30. A

RATE YOURSELF!!!
for the 1997-98 ALGEBRA COURSE 1 CONTEST

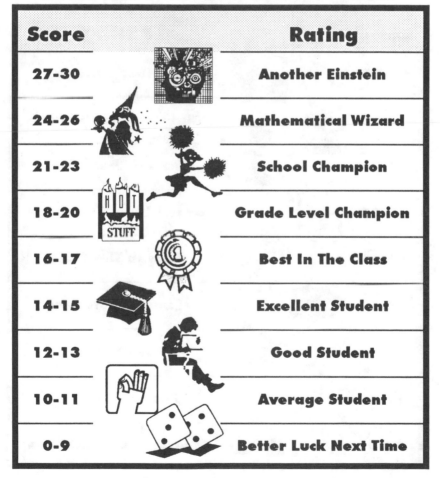

Score	Rating
27-30	Another Einstein
24-26	Mathematical Wizard
21-23	School Champion
18-20	Grade Level Champion
16-17	Best In The Class
14-15	Excellent Student
12-13	Good Student
10-11	Average Student
0-9	Better Luck Next Time

ANSWERS, 1998-99 Algebra Course 1 Contest

1. B	7. C	13. A	19. C	25. D
2. D	8. A	14. B	20. A	26. B
3. B	9. B	15. B	21. D	27. C
4. A	10. A	16. D	22. C	28. B
5. C	11. D	17. C	23. D	29. A
6. C	12. D	18. A	24. A	30. D

RATE YOURSELF!!!
for the 1998-99 ALGEBRA COURSE 1 CONTEST

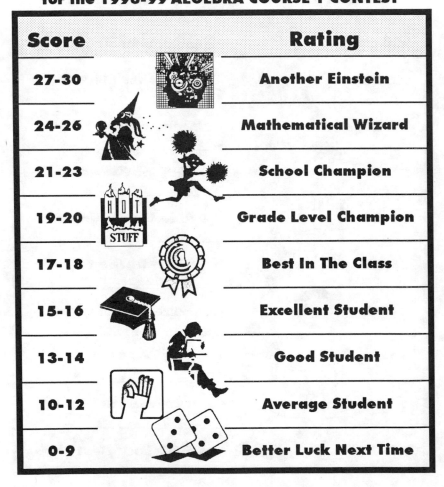

Score	Rating
27-30	Another Einstein
24-26	Mathematical Wizard
21-23	School Champion
19-20	Grade Level Champion
17-18	Best In The Class
15-16	Excellent Student
13-14	Good Student
10-12	Average Student
0-9	Better Luck Next Time

ANSWERS, 1999-00 Algebra Course 1 Contest

1. A	7. C	13. B	19. D	25. D
2. A	8. C	14. B	20. B	26. D
3. A	9. A	15. D	21. A	27. A
4. B	10. A	16. C	22. B	28. C
5. A	11. D	17. D	23. C	29. B
6. B	12. D	18. C	24. D	30. D

RATE YOURSELF!!!
for the 1999-00 ALGEBRA COURSE 1 CONTEST

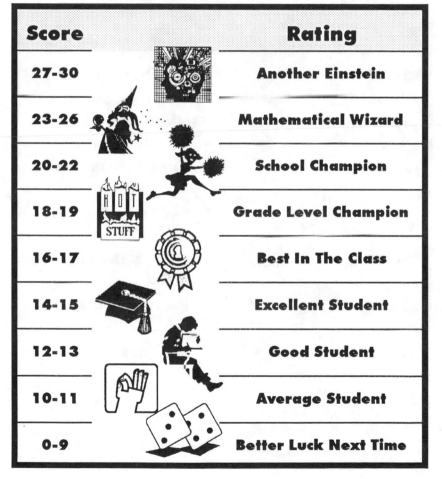

Score	Rating
27-30	Another Einstein
23-26	Mathematical Wizard
20-22	School Champion
18-19	Grade Level Champion
16-17	Best In The Class
14-15	Excellent Student
12-13	Good Student
10-11	Average Student
0-9	Better Luck Next Time

ANSWERS, 2000-01 Algebra Course 1 Contest

1. A	7. D	13. D	19. D	25. B
2. B	8. C	14. D	20. A	26. D
3. B	9. A	15. B	21. A	27. C
4. D	10. D	16. C	22. A	28. A
5. C	11. B	17. C	23. C	29. D
6. C	12. A	18. B	24. A	30. B

RATE YOURSELF!!!
for the 2000-01 ALGEBRA COURSE 1 CONTEST

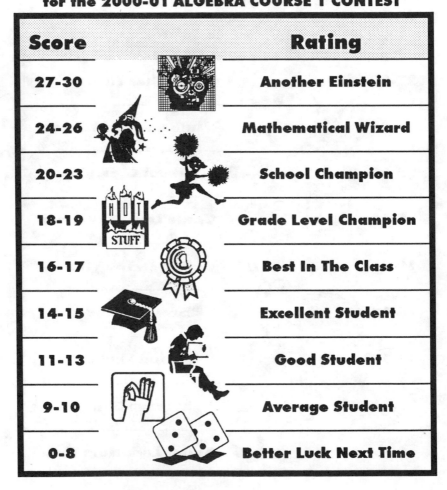

Score	Rating
27-30	Another Einstein
24-26	Mathematical Wizard
20-23	School Champion
18-19	Grade Level Champion
16-17	Best In The Class
14-15	Excellent Student
11-13	Good Student
9-10	Average Student
0-8	Better Luck Next Time

Math League Contest Books
4th Grade Through High School Levels

Written by Steven R. Conrad and Daniel Flegler, recipients of President Reagan's 1985 Presidential Awards for Excellence in Mathematics Teaching, each book provides schools and students with:

- Easy-to-use format designed for a 30-minute period
- Problems ranging from straightforward to challenging

Use the form below (or a copy) to order your books

Name: _____

Address: _____

City: _____ State: _____ Zip: _____
                        (or Province)   *(or Postal Code)*

Available Titles	# of Copies	Cost
Math Contests—Grades 4, 5, 6	($12.95 ea., $19.95 Canadian)	
Volume 1: 1979-80 through 1985-86	_____	_____
Volume 2: 1986-87 through 1990-91	_____	_____
Volume 3: 1991-92 through 1995-96	_____	_____
Volume 4: 1996-97 through 2000-01	_____	_____
Math Contests—Grades 7 & 8 ‡	‡(Vols. 3 & 4 include Alg. Course I)	
Volume 1: 1977-78 through 1981-82	_____	_____
Volume 2: 1982-83 through 1990-91	_____	_____
Volume 3: 1991-92 through 1995-96	_____	_____
Volume 4: 1996-97 through 2000-01	_____	_____
Math Contests—High School		
Volume 1: 1977-78 through 1981-82	_____	_____
Volume 2: 1982-83 through 1990-91	_____	_____
Volume 3: 1991-92 through 1995-96	_____	_____
Volume 4: 1996-97 through 2000-01	_____	_____
Shipping and Handling	$3 ($5 Canadian)	

Please allow 4-6 weeks for delivery   Total: $_____

☐ Check or Purchase Order Enclosed; **or**

☐ Visa / MstrCrd / Discvr # _____

☐ Exp. Date_____ Signature _____

Mail your order with payment to:

Math League Press, P.O. Box 17, Tenafly, NJ USA 07670-0017

(or order on the web at www.mathleague.com)

Phone: (201) 568-6328 • Fax: (201) 816-0125